Shortcut Calculus I

Shortcut
Calculus

Anaxos, Inc.

PUBLISHING

New York • Chicago

This publication is designed to provide accurate and authoritative information in regard to the subject matter covered. It is sold with the understanding that the publisher is not engaged in rendering legal, accounting, or other professional service. If legal advice or other expert assistance is required, the services of a competent professional should be sought.

Editorial Director: Jennifer Farthing
Senior Editor: Tonya Lobato
Production Artist: Baldur Gudbjornsson
Cover Designer: Carly Schnur

© 2006 by Kaplan, Inc.

Published by Kaplan Publishing, a division of Kaplan, Inc.
888 Seventh Ave.
New York, NY 10106

Printed in the United States of America

June 2006

10 9 8 7 6 5 4 3 2 1

ISBN-13: 978-1-4195-4163-3
ISBN-10:   1-4195-4163-6

Kaplan Publishing books are available at special quantity discounts to use for sales promotions, employee premiums, or educational purposes. Please call our Special Sales Department to order or for more information at 800-621-9621, ext. 4444, e-mail kaplanpubsales@kaplan.com, or write to Kaplan Publishing, 30 South Wacker Drive, Suite 2500, Chicago, IL 60606-7481.

TABLE OF CONTENTS

CHAPTER 4: LINEAR EQUATIONS

CHAPTER 5: TRIGONOMETRIC FUNCTIONS

CHAPTER 6: LIMITS AND CONTINUITY

CHAPTER 7: DERIVATIVES

CHAPTER 8: APPLICATIONS OF DERIVATIVES

CHAPTER 9: INTEGRATION

CHAPTER 10: APPLICATIONS OF THE DEFINITE INTEGRAL

CHAPTER 11: MAX-MIN AND OTHER RATES

CHAPTER 12: REAL-WORLD APPLICATIONS

Introduction

If you have picked up this book, you might be one of the many people who thinks calculus is an intimidating topic, but who needs to have basic knowledge of the subject. Maybe you're a student gearing up for a standardized test that includes calculus. Or maybe you're a first-time calculus student who needs some additional guidance in a high school or college class.

If you are one of these people, this book is designed just for you! *Shortcut Calculus* offers an easy-to-understand approach that will guide you through the maze of problems and proofs that comprise basic calculus. With over 200 step-by-step examples and practice questions, you'll be well on your way to feeling confident and at-ease with this challenging subject.

Shortcut Calculus begins with a review of common algebra and trigonometry topics that are the foundation for calculus—often referred to as "pre-calculus." Next comes a step-by-step introduction to the heart of the subject: limits, derivatives, and integrals. Throughout, our discussion includes examples of how to apply concepts to concrete, real-world problems—essential for bringing calculus out of the realm of the abstract and into everyday light.

To use the book to the fullest advantage, start by taking the Diagnostic Quiz. Following the quiz, the Diagnostic Correlation Chart and detailed answer explanations will help you identify your weak areas and will direct you to the appropriate chapters in the book for review. Depending on your needs, you may choose to skip directly ahead to those chapters or to work through the whole book from beginning to end.

If you have enough time to do so, we recommend that you work through the entire book, because each chapter builds on information presented in previous chapters. It would be to your advantage to review all the concepts and practice all the problems. For example, to understand integrals, you need to understand derivatives, and to understand derivatives, you need to understand limits. Similarly, evaluating derivatives of trigonometric functions requires that you first understand trigonometric functions.

Each core chapter is structured to identify key concepts and outline steps that will help you to solve the most common types of questions. You'll learn how to apply those steps to real problems. As you wend your way to a solution, detailed explanations will walk you through problem-solving techniques and useful strategies. Each chapter concludes with a 10-problem quiz, which will help you to evaluate yourself and apply your understanding. In-depth explanatory solutions are provided for the quiz problems as well, so that you can check your work and target any areas where you may still need some review.

By the time you reach the end of *Shortcut Calculus*, we're confident that you will see calculus in a whole new light—and you'll be amazed that it took such a short time to get from where you started to a clear understanding of the basics. You will be well on your way to mastering the essential skills and concepts that you need to succeed in calculus.

Good luck—and enjoy the shortcut!

Diagnostic Test

Before you begin using this book, take this 30-question diagnostic test. The questions on this test cover the topics you will encounter in *Shortcut Calculus*. Your performance on the diagnostic will provide you with a general picture of the subjects in which you are strong and the topics you need to spend time reviewing. You can use this information to tailor your approach to the chapters in this book. Ideally, you should read all the chapters, but if you're pressed for time, you can start with the chapters and subjects you really need to study.

Be sure to read the explanations for all of the questions, including those you answered correctly. Even if you got the problem right, reading another person's answer can give you insights that will prove helpful.

Good luck on the diagnostic test!

Diagnostic Test

Directions: Solve the following problems. After examining the choices, decide which one is the best.

The following rules apply to this test:

(1) The domain of a function f is the set of all real numbers x for which $f(x)$ is a real number, unless otherwise specified.

(2) The inverse of a trigonometric function f may be indicated using the inverse function notation f^{-1} or with the prefix "arc" (e.g., $\sin^{-1} x = \arcsin x$).

1. Express the graph below in interval notation:

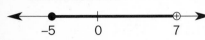

   (A) (–5,7)

   (B) [–5,7]

   (C) (–5,7]

   (D) [–5,7)

   (E) [–5,∞)

2. Express $0 < x \le 9$ in interval notation:

   (A) (0,9)

   (B) [0,9]

   (C) (0,9]

   (D) [0,9)

   (E) (–∞,9]

3. Simplify $|17 - 22|$

   (A) 5

   (B) –5

   (C) 39

   (D) –39

   (E) none of the above

4. Solve $|x + 14| = 8$

   (A) 22 and 6

   (B) –22 and 6

   (C) 22 and –6

   (D) –22 and –6

   (E) none of the above

5. Graph $f(x) = x^2 + 4x - 2$.

(A) 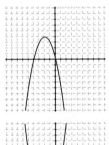 (B) (C)

(D) 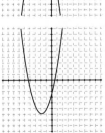 (E)

6. Solve $f(x) = 3x^3 + 6x^2 - 9x$

   (A) $\{-3\}$

   (B) $\{0\}$

   (C) $\{1\}$

   (D) $\{-3,0,1\}$

   (E) more information is needed to solve this problem

7. Find the slope of a line that passes through the points $(-12,-1)$ and $(7,5)$.

   (A) $\dfrac{19}{6}$

   (B) $\dfrac{6}{19}$

   (C) $-\dfrac{6}{19}$

   (D) 0

   (E) the slope is undefined

8. Write the equation of a line that contains the point $(-1,6)$ and has a slope of 5.

   (A) $y = 5x - 1$

   (B) $y = 5x + 6$

   (C) $y = 5x + 1$

   (D) $y = 5x - 6$

   (E) $y = 5x + 11$

9. Find the value of $\cos\dfrac{15\pi}{4}$.

   (A) $\dfrac{\sqrt{2}}{2}$

   (B) $\dfrac{1}{2}$

   (C) $\dfrac{\sqrt{3}}{2}$

   (D) 0

   (E) 1

10. Solve $\sqrt{2}\csc x - 2 = 0$.

   (A) $\dfrac{\pi}{6}$

   (B) $\dfrac{\pi}{4}$

   (C) $\dfrac{\pi}{3}$

   (D) $\dfrac{\pi}{2}$

   (E) π

11. Evaluate $\lim\limits_{x \to 1} \dfrac{2x^2 + x - 3}{1 - x^2}$.

   (A) $-\dfrac{5}{2}$

   (B) 0

   (C) $\dfrac{1}{2}$

   (D) $-\dfrac{3}{2}$

   (E) The limit does not exist.

12. Evaluate $\lim\limits_{x\to\infty} \dfrac{x^3 + 3x^5 + 6x}{2x^5 + 3x^2 + 5}$.

  (A) 0

  (B) $-\dfrac{1}{2}$

  (C) $\dfrac{2}{3}$

  (D) $\dfrac{3}{2}$

  (E) The limit does not exist.

13. Evaluate $\lim\limits_{x\to 0} \dfrac{\sin^2(4x)}{x^2}$.

  (A) $\dfrac{1}{4}$

  (B) 0

  (C) 16

  (D) 4

  (E) The limit does not exist.

14. If $c(x) = 2000 + 8.6x + 0.5x^2$, then $c'(300) =$

  (A) 313.6

  (B) 308.6

  (C) 300.0

  (D) 297.2

  (E) 200.0

15. If $f(x) = \cos^2(\sin(2x))$, then $f'(\frac{\pi}{8}) =$

  (A) $1 + \sqrt{2}\cos(\frac{\sqrt{2}}{2})$

  (B) $\sqrt{2}\sin(\frac{\sqrt{2}}{2})$

  (C) $\cos(\sin(\frac{\sqrt{2}}{2}))$

  (D) $\dfrac{\sqrt{2}}{2}\sin(\frac{\sqrt{2}}{2})$

  (E) $-\sqrt{2}\cos(\frac{\sqrt{2}}{2})$

16. The slope of the tangent to the curve $2y^2x^3 - 5x^2y = 18$ at the point $(1,2)$ is

  (A) $-\dfrac{4}{5}$

  (B) $-\dfrac{3}{2}$

  (C) 0

  (D) $-\dfrac{4}{3}$

  (E) -1

17. An equation of the line tangent to the graph of $y = 2\sin(2x + \frac{3\pi}{4})$ at $x = \frac{\pi}{8}$ is

  (A) $y - \frac{\pi}{2} = -4x$

  (B) $y + \frac{\pi}{4} = \frac{3}{4}x$

  (C) $y = 4x + \frac{\pi}{2}$

  (D) $y + \frac{\pi}{2} = 4x$

  (E) $2y = -4x + \pi$

18. What are the values for which the function $f(x) = \frac{2}{3}x^3 - x^2 - 4x + 3$ is increasing?

(A) $-1 < x < 2$

(B) $x < -1$

(C) $x < -1$ and $x > 2$

(D) $0 < x < 2$

(E) $x > -1$

19. The graph of $y = x^3 + 21x^2 - x + 1$ is concave down for

(A) $x < -7$

(B) $-7 < x < 7$

(C) all x

(D) $x > 7$

(E) $x < -7$ and $x > 7$

20. Evaluate $\int (\sin(2x) + 5\tan^2 x \csc^2 x)dx$.

(A) $-\frac{1}{2}\cos(2x) + 5\tan x + C$

(B) $\cos(2x) - 5\cot x + C$

(C) $-\cos(2x) + 5\sec x \tan x + C$

(D) $-\frac{1}{2}\cos(2x) - 5\cot x + C$

(E) $\frac{1}{2}\sin(2x) + 5\tan x + C$

21. Evaluate $\int_{-1}^{2}\left(3x^2 - 4x + 2\right)dx$.

(A) -2

(B) 14

(C) 9

(D) 18

(E) 21

22. Find the derivative of $f(x) = \int_{0}^{x^2} \cos(t^2)dt$.

(A) $\cos\left(x^4\right)$

(B) $\cos\left(x^2\right)$

(C) $2x\cos\left(x^2\right)$

(D) $x\cos\left(x^4\right)$

(E) $2x\cos\left(x^4\right)$

23. The area bounded by the curves $y = x^2 + 4$ and $y = -2x + 1$ between $x = -2$ and $x = 5$ equals

(A) 86.500

(B) 86.425

(C) 86.333

(D) 86.125

(E) 86.000

24. The area of the region enclosed by the graph of $y = 2x^2 + 1$ and the line $y = 2x + 5$ is

(A) 5

(B) 3

(C) $\frac{7}{2}$

(D) 9

(E) 1

25. If the region enclosed by the x-axis, the line $x = 1$, the line $x = 3$, and the curve $y = x^{\frac{3}{2}}$ is revolved around the x-axis, the volume of the solid is

(A) 20π

(B) $\frac{25}{2}\pi$

(C) 21π

(D) 4

(E) $\frac{100\pi}{3}$

26. Consider the function
$$f(x) = \begin{cases} x^2 \text{ if } 0 \leq x < 1 \\ 0 \text{ if } 1 \leq x \leq 2 \end{cases}.$$
Which of the following is true?

(A) f attains an absolute maximum value of 1.

(B) f attains an absolute minimum value of 0.

(C) f attains an absolute maximum value somewhere on the interval $[0, 2]$.

(D) f does not attain an absolute minimum value.

(E) Both (A) and (C)

27. A particle moves along the x-axis so that its velocity at any time $t > 0$ is given by $v(t) = 5t^2 - 4t + 7$. At what point does the particle achieve a minimum velocity?

(A) 0

(B) 5

(C) 4

(D) $\frac{2}{5}$

(E) 7

28. A gun is fired vertically upward from a position 100 feet above ground at an initial velocity of 400 feet per second. Determine the maximum height of the projectile. The acceleration of gravity is $-32\dfrac{ft}{sec^2}$.

    (A) 3000 feet
    (B) 2600 feet
    (C) 2200 feet
    (D) 1800 feet
    (E) 1400 feet

29. A 13-foot ladder is leaning against a 20-foot vertical wall when it begins to slide down the wall. During this sliding process, the bottom of the ladder is sliding away from the bottom of the wall at a rate of $\frac{1}{2}$ a foot per second. Determine the rate at which the top of the ladder is sliding down the vertical wall when the tip of the ladder is exactly 5 feet above the ground.

    (A) $-\frac{6}{5}$ feet per second
    (B) $\frac{5}{6}$ feet per second
    (C) $-\frac{12}{13}$ feet per second
    (D) -2 feet per second
    (E) There is not enough information here to solve this problem.

30. A particle moves along the x-axis so that its velocity at any time $t > 0$ is given by $v(t) = 5t^2 - 4t + 7$. The position of the particle, $x(t)$, is 8 for $t = 3$. What is the total distance traveled by the particle from time $t = 0$ until the time $t = 2$?

    (A) $\frac{3}{2}$
    (B) 1.9
    (C) $\frac{5}{4}$
    (D) 2
    (E) $\frac{58}{3}$

Answers and Explanations

1. D

The graph represents the range of numbers between –5 and 7, including –5 but *not* including 7. The solid circle (●) indicates that –5 is included in the range and the open circle (○) indicates that 7 is not. In interval notation, we express ● as a bracket "[" and ○ as a parenthesis ")". Therefore, the interval notation is [–5,7).

2. C

The inequality expresses the range of possible values for x: the value of x is greater than 0 but less than or equal to 9. The inequality symbol < indicates that 0 is not a possible value for x, so 0 is not included in the interval. The ≤ symbol indicates that 9 is a possible value for x, so 9 is included in the interval. We use [or] to denote a value that is included in the interval and (or) to denote a value that is *not* included. Therefore, the interval notation for the given inequality is (0,9].

3. A

First, we need to simplify the expression within the absolute value symbols, | and |.

$$|17 - 22| = |-5|$$

The absolute value of a number is always positive. Therefore, $|-5| = 5$.

4. D

To solve for $|x + 14| = 8$, we need to solve for $x + 14 = 8$ and $x + 14 = -8$.

$$
\begin{array}{ccc}
x + 14 = 8 & & x + 14 = -8 \\
x + 14 - 14 = 8 - 14 & \text{and} & x + 14 - 14 = -8 - 14 \\
x = -6 & & x = -22
\end{array}
$$

We can verify our solutions by plugging them into the original equation.

$$|x + 14| = |-6 + 14| = |8| = 8$$
$$|x + 14| = |-22 + 14| = |-8| = 8$$

5. E

To graph $f(x) = x^2 + 4x - 2$, we make a table of values for x and $f(x)$. First, we choose some values for x and plug them into the function to solve for $f(x)$.

x	$f(x)$	$(x, f(x))$
-2	$(-2)^2 + 4(-2) - 2 = -6$	$(-2, -6)$
-1	$(-1)^2 + 4(-1) - 2 = -5$	$(-1, -5)$
0	$(0)^2 + 4(0) - 2 = -2$	$(0, -2)$
1	$(1)^2 + 4(1) - 2 = 3$	$(1, 3)$
2	$(2)^2 + 4(2) - 2 = 10$	$(2, 10)$

Next, we plot the ordered pairs $(x, f(x))$ on a x, y plane and draw a smooth curve through the points to get the graph of $f(x) = x^2 + 4x - 2$.

6. D

To solve the polynomial function, we set the polynomial equation equal to zero and then factor.

$$3x^3 + 6x^2 - 9x = 0$$
$$3x\left(x^2 + 2x - 3\right) = 0$$
$$3x(x + 3)(x - 1) = 0$$

Next, we solve for x by setting each of the factors equal to 0.

$$3x = 0 \quad \text{and} \quad x + 3 = 0 \quad \text{and} \quad x - 1 = 0$$
$$x = 0 \qquad\qquad x = -3 \qquad\qquad x = 1$$

Therefore, our solution set is $\{-3, 0, 1\}$.

Verify that each of these values for x results in $f(x) = 0$.

$$3(-3)^3 + 6(-3)^2 - 9(-3) = -81 + 54 + 27 = 0$$
$$3(0)^3 + 6(0)^2 - 9(0) = 0 + 0 + 0 = 0$$
$$3(1)^3 + 6(1)^2 - 9(1) = 3 + 6 - 9 = 0$$

7. B

The formula for the slope of a line is $m = \dfrac{y_2 - y_1}{x_2 - x_1}$, where m is the slope and (x_1, y_1) and (x_2, y_2) are two points on the line. We are given $(-12, -1)$ and $(7, 5)$ as two points on the line, so $x_1 = -12$, $y_1 = -1$, $x_2 = 7$, and $y_2 = 5$.

The slope of the line is $m = \dfrac{5 - (-1)}{7 - (-12)} = \dfrac{5 + 1}{7 + 12} = \dfrac{6}{19}$.

8. E

We know the slope and a point on the line, so we can write the point-slope equation of the line. The point-slope form of a line is $y - y_0 = m(x - x_0)$, where m is the slope of the line and (x_0, y_0) is a point on the line. We know that $m = 5$ and (x_0, y_0) is $(-1, 6)$; therefore

$$
\begin{aligned}
y - y_0 &= m(x - x_0) \\
y - 6 &= 5(x - (-1)) \\
y - 6 &= 5(x + 1) \\
y - 6 &= 5x + 5 \\
y - 6 + 6 &= 5x + 5 + 6 \\
y &= 5x + 11
\end{aligned}
$$

9. A

An angle coterminal to $\dfrac{15\pi}{4}$ would be $\dfrac{15\pi}{4} - 2\pi = \dfrac{15\pi}{4} - \dfrac{8\pi}{4} = \dfrac{7\pi}{4}$.

An angle coterminal to $\dfrac{7\pi}{4}$ would be $\dfrac{7\pi}{4} - 2\pi = \dfrac{7\pi}{4} - \dfrac{8\pi}{4} = -\dfrac{\pi}{4}$.

If we construct a right triangle on an xy coordinate system with an angle of $-\dfrac{\pi}{4}$ (which is $-45°$) and a hypotenuse of 1, we see that the legs of the right triangle are both $\dfrac{\sqrt{2}}{2}$.

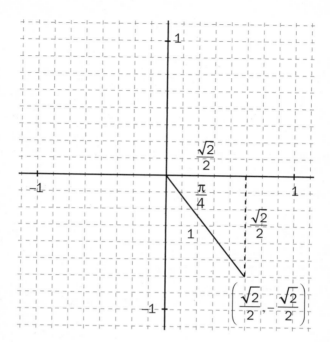

Therefore, $\cos-\dfrac{\pi}{4} = \dfrac{\text{adjacent}}{\text{hypotenuse}} = \dfrac{\sqrt{2}}{2}$.

10. B

We solve for x by first solving for $\csc x$:

$$\sqrt{2} \csc x - 2 = 0$$

$$\sqrt{2} \csc x = 2$$

$$\csc x = \dfrac{2}{\sqrt{2}}$$

Since $\csc x = \dfrac{1}{\sin x}$, we can solve for $\sin x$:

$$\dfrac{1}{\sin x} = \dfrac{2}{\sqrt{2}}$$

$$\sin x = \dfrac{\sqrt{2}}{2}$$

We should know, as a memorized fact, that $\sin \dfrac{\pi}{4} = \dfrac{\sqrt{2}}{2}$, so $x = \dfrac{\pi}{4}$.

11. A

If we plug in the value $x = 1$ we get

$$\lim_{x \to 1} \frac{2x^2 + x - 3}{1 - x^2} = \lim_{x \to 1} \frac{2(1)^2 + (1) - 3}{(1 - (1)^2)} = \lim_{x \to 1} \frac{2 + 1 - 3}{1 - 1} = \frac{0}{0} \, ,$$

which is an indeterminate value.

However, by factoring the numerator and the denominator, we can reduce the argument and easily plug in the value of $x = 1$ to evaluate the limit:

$$\lim_{x \to 1} \frac{2x^2 + x - 3}{1 - x^2} = \lim_{x \to 1} \frac{(2x + 3)(x - 1)}{(1 + x)(1 - x)} = \lim_{x \to 1} \frac{-(2x + 3)(1 - x)}{(1 + x)(1 - x)} = \lim_{x \to 1} \frac{-(2x + 3)}{1 + x} = \frac{-5}{2} \, .$$

12. D

The term of highest degree in the numerator, $3x^5$, is controlling the behavior of the numerator; therefore, the numerator of the function "looks like" $3x^5$. Similarly, the term of highest degree in the denominator, $2x^5$, is controlling the behavior of the denominator; therefore, the denominator "looks like" $2x^5$.

Because the degree of the numerator polynomial is equal to the degree of the denominator polynomial, the limit as x approaches infinity is simply the ratio of the coefficients of the highest-order terms:

$$\lim_{x \to \infty} \frac{x^3 + 3x^5 + 6x}{2x^5 + 3x^2 + 5} = \lim_{x \to \infty} \frac{3x^5}{2x^5} = \frac{3}{2} \, .$$

13. C

For this problem, the limit can be rewritten:

$$\lim_{x \to 0} \frac{\sin^2(4x)}{x^2} = \lim_{x \to 0} \left(\frac{\sin 4x}{x} \right)^2 \, .$$

You should know that $\lim_{x \to \infty} \frac{\sin x}{x} = 1$. This rule applies as long as the variable in the numerator is the same as the variable in the denominator.

To get $\frac{\sin 4x}{4}$ into the form $\frac{\sin x}{x}$, we multiply by the numerator and the denominator by 4:

$$\lim_{x \to 0} \left(\frac{\sin 4x}{x} \right)^2 = \lim_{x \to 0} \left(\frac{4}{4} \times \frac{\sin 4x}{x} \right)^2 = \lim_{x \to 0} \left(4 \times \frac{\sin 4x}{4x} \right)^2 \, .$$

Next we factor out the constant, which is squared:

$$\lim_{x\to 0}\left(4\times\frac{\sin 4x}{4x}\right)^2 = 16\times\lim_{x\to 0}\left(\frac{\sin 4x}{4x}\right)^2 = 16\times 1 = 16$$

Note that as $x\to 0$, $4x\to 0$, so the limit is still 1.

14. B

The derivative of any constant is zero, so $\frac{d}{dx}(2000)=0$.

Using the power rule, we find that:

$$\frac{d}{dx}(8.6x) = (8.6)(1)(x^{1-1}) = (8.6)(1)(x^0) = 8.6 \text{ and}$$

$$\frac{d}{dx}(0.5x^2) = (0.5)(2)(x^{2-1}) = (0.5)(2)(x) = 1.0x$$

Therefore, $c'(x) = 0 + 8.6 + 1.0x$ and $c'(300) = 0 + 8.6 + 1.0(300) = 308.6$.

15. E

In this problem, it helps to rewrite f as $f(x) = [\cos(\sin(2x))]^2$ and then we have to use the chain rule three times.

We start by differentiating the outermost function:

$$\frac{d}{dx}[\cos(\sin(2x))]^2 = 2\times[\cos(\sin(2x))]^{2-1} = 2[\cos(\sin 2x))].$$

Then we differentiate the inner function $[\cos(\sin(2x))]$. However, since $[\cos(\sin(2x))]$ is itself a composite function, we use the chain rule a second time to differentiate $[\cos(\sin(2x))]$ and its inner function $\sin(2x)$:

$$\frac{d}{dx}[\cos(\sin(2x))] = -\sin(\sin(2x))\times\cos(2x).$$

Notice that $\sin(2x)$ is also a composite function, so we apply the chain rule a third time. We have already differentiated $\sin(2x)$ above, so we now need to differentiate its inner function, $2x$:

$$\frac{d}{dx}(2x) = 2.$$

By applying the chain rule (three times) we get

$$f'(x) = 2[\cos(\sin(2x))] \times (-\sin(\sin(2x))) \times \cos(2x) \times 2$$

$$= -4\cos(\sin(2x)) \times \sin^2(2x) \times \cos(2x)$$

Now by plugging in the value of $x = \frac{\pi}{8}$ and using the fact that $\sin\frac{\pi}{4} = \cos\frac{\pi}{4} = \frac{\sqrt{2}}{2}$, we get $f'\left(\frac{\pi}{8}\right) = -\sqrt{2}\cos\left(\frac{\sqrt{2}}{2}\right)$.

16. D

Differentiate the expression term-by-term, keeping the product rule in mind and attaching a term dx or dy each time we differentiate x or y, respectively. Then solve for $\frac{dy}{dx}$. We get

$$(2y^2 x^3) - (5x^2 y) = 18$$

$$(4yx^3 dy + 6y^2 x^2 dx) - (10xy \ dx + 5x^2 dy) = 0$$

$$4yx^3 dy - 5x^2 dy = 10xy \ dx - 6y^2 x^2 dx$$

$$(4yx^3 - 5x^2)dy = (10xy - 6y^2 x^2)dx$$

$$\frac{dy}{dx} = \frac{(10xy - 6y^2 x^2)}{(4yx^3 - 5x^2)}.$$

Now evaluate $\frac{dy}{dx}$ at the point (1,2) to get the slope of the tangent line the curve. We get

$$\left.\frac{dy}{dx}\right|_{(1,2)} = \frac{[10(1)(2) - 6(2)^2 (1)^2]}{[4(2)(1)^3 - 5(1)^2]} = -\frac{4}{3}$$

17. D

To get the equation of the tangent line, we need to know the slope of the tangent line and a point on the tangent line.

Using the chain rule, $y'\left(\frac{\pi}{8}\right) = 2\cos\left(2x + \frac{3\pi}{4}\right) \cdot 2 = 4\cos\left(2x + \frac{3\pi}{4}\right) = 4\cos\left(\frac{\pi}{4} + \frac{3\pi}{4}\right) = 4.$

At $x = \frac{\pi}{8}$, $y = 2\sin\left(2 \cdot \frac{\pi}{8} + \frac{3\pi}{4}\right) = 2\sin\left(\frac{\pi}{4} + \frac{3\pi}{4}\right) = 2(\sin\pi) = 2(0) = 0.$

Therefore $\left(\frac{\pi}{8}, 0\right)$ is a point on the tangent line.

Using the point-slope form we conclude that the equation of the tangent line is $y + \dfrac{\pi}{2} = 4x$.

18. C

The function f is increasing for all x such that $f'(x) > 0$. By direct calculation,

$$
\begin{aligned}
f'(x) &= 3 \cdot \frac{2}{3}\left(x^{3-1}\right) - 2\left(x^{2-1}\right) - 4\left(x^{1-1}\right) + 0 \\
&= 2x^2 - 2x - 4 = 2\left(x^2 - x - 2\right) \\
&= 2(x-2)(x+1)
\end{aligned}
$$

Therefore, $f'(x) > 0$ when $(x - 2)$ and $(x + 1)$ are either both positive or both negative. This occurs when $x < -1$ and $x > 2$.

19. A

The function f is concave down for all x such that $f''(x) < 0$. By direct calculation,

$$
\begin{aligned}
y' &= 3\left(x^{3-1}\right) + 2 \cdot 21\left(x^{2-1}\right) - x^{1-1} + 0 \\
&= 3x^2 + 42x - 1
\end{aligned}
$$

and

$$
\begin{aligned}
y'' &= 2 \cdot 3\left(x^{2-1}\right) + 42\left(x^{1-1}\right) - 0 \\
&= 6x + 42
\end{aligned}
$$

So f is concave down when $6x + 42 < 0$, that is, when $x < -7$.

20. A

First, we note that an antiderivative of $\sin(2x)$ is $-\dfrac{1}{2}\cos(2x)$.

Next we have $5 \tan^2 x \csc^2 x = 5\dfrac{\sin^2 x}{\cos^2 x} \cdot \dfrac{1}{\sin^2 x} = 5\dfrac{1}{\cos^2 x} = 5\sec^2 x$.

An antiderivative of $\sec^2 x$ is $\tan x$.

21. C

To evaluate this integral, we can first rewrite it as $\int_{-1}^{2} 3x^2 dx - \int_{-1}^{2} 4x dx + \int_{-1}^{2} 2dx$.

Carefully taking the antiderivative of the integrand, and then plugging in the limits of integration will yield the correct answer.

The *derivative* of x^3 is $3x^2$, so $\int_{-1}^{2} 3x^2 dx = x^3 \Big|_{-1}^{2} = (2)^3 - (-1)^3 = 9$.

The *derivative* of $2x^2$ is $4x$, so $\int_{-1}^{2} 4x dx = 2x^2 \Big|_{-1}^{2} = 2(2)^2 - 2(-1)^2 = 6$.

The *derivative* of $2x$ is 2, so $\int_{-1}^{2} 2dx = 2x \Big|_{-1}^{2} = 2(2) - 2(-1) = 6$.

Therefore, the integral is $\quad 9 - 6 + 6 = 9$.

22. E

Set $f(x) = \int_0^x \cos(t^2) dt$.

Therefore, $f(x^2) = \int_0^{x^2} \cos(t^2) dt$.

The Second Fundamental Theorem of Calculus states that $\dfrac{d}{dx} \int_a^x f(t)\, dt = f(x)$.

Therefore, $f'(x) = \cos(x^2)$ and $f'(x^2) = \cos\left((x^2)\right)^2$.

Now applying the chain rule we see that

$$\frac{d}{dx}\left(\int_0^{x^2} \cos(t^2)\, dt\right) = \frac{d}{dx} f\left(x^2\right)$$

$$= f'\left(x^2\right) \cdot 2x$$

$$= \cos\left(\left(x^2\right)^2\right) \cdot 2x$$

$$= 2x \cos\left(x^4\right)$$

23. C

When we graph the two curves $y = x^2 + 4$ and $y = -2x + 1$, we see that $y = x^2 + 4$ is always above $y = -2x + 1$ on the interval $[-2, 5]$.

To compute the area between two curves between $x = a$ and $x = b$, we evaluate the integral

$$\int_a^b \text{function on top} - \text{function on bottom } dx \ .$$

Therefore, the area in question is found by computing the value of

$$\int_{-2}^{5} (x^2 + 4) - (-2x + 1)\, dx = \int_{-2}^{5} (x^2 + 2x + 3)\, dx$$

$$= \left(\frac{x^3}{3} + x^2 + 3x \right)\Bigg|_{-2}^{5}$$

$$= \left[\frac{(5)^3}{3} + (5)^2 + 3(5) \right] - \left[\frac{(-2)^3}{3} + (-2)^2 + 3(-2) \right]$$

$$= \frac{259}{3}$$

The final answer is $\frac{259}{3}$, which is approximately 86.3333.

24. D

To calculate the area between the curves $y = 2x^2 + 1$ and $y = 2x + 5$, we must evaluate the integral $\int_a^b (2x + 5) - (2x^2 + 1)\, dx$.

To determine which values to use for a and b as the limits of the integral, we calculate the x values where the two curves intersect by solving the equation $2x^2 + 1 = 2x + 5$.

We can rewrite the equation as $2x^2 - 2x - 4 = 0$ and factor to get $x = 2$ and $x = -1$.

Set $a = -1, b = 2$. The enclosed area, A, is thus given by the equation

$$A = \int_{-1}^{2} (2x + 5) - (2x^2 + 1)\ dx = \int_{-1}^{2} -2x^2 + 2x + 4\ dx = \left[\frac{-2x^3}{3} + x^2 + 4x \right]\Bigg|_{-1}^{2} = -6 + 15 = 9.$$

25. A

When we graph the curve and revolve it around the x-axis, we find that the cross section of the resulting solid is a disk. Because the solid is revolving around the x-axis, the radius of the disk will be the value of y.

We will use the term $\pi[f(x)]^2$ to measure the area of a cross-sectional disk of the solid where $f(x)$ = radius of cross section given by the function.

The term $\int_a^b dx$, where a and b are the left and right bounds of the surface of revolution, adds up all the disks to give the entire volume of the solid. Therefore, the volume of the solid is

$$V = \int_a^b \pi[f(x)]^2 \, dx = \int_1^3 \pi[x^{\frac{3}{2}}]^2 \, dx = \int_1^3 \pi x^3 \, dx = \frac{\pi x^4}{4}\Big|_1^3 = \pi\left[\frac{3^4}{4} - \frac{1}{4}\right] = 20\pi$$

26. B

This is a problem in which visualizing the graph of the given function proves extremely helpful. The function is defined piecewise, and each piece is simple.

We know that the graph of $f(x) = x^2$ is a parabola for all real numbers of x. On the interval $[0,1)$, the graph is the right half of the parabola with $f(x)$ getting closer to 1 as x gets closer to 1. However, at $x = 1$, $f(x) = 0$, so the function jumps down to 0. In fact, on the interval $[1,2]$, we simply get a line along the x-axis.

Therefore, the highest value on the graph is almost 1 and the lowest value is 0. At first glance, it may seem that choices (A), (B), and (C) are all correct, but the key word here is "attains." Choice (A) is incorrect because the maximum value of 1 is never attained by a particular x, and similar reasoning eliminates (C). But f does attain its absolute minimum value of 0, so (B) must be the correct choice. Visualize the graphs of functions whenever possible.

27. D

When $v'(t) = 0$, the velocity is changing direction. Solving $v'(t) = 0$ gives us the point(s) at which the velocity changes from increasing to decreasing or from decreasing to increasing. The second derivative test tells us whether the point is a minimum or maximum. If $v''(t) > 0$, the point is a minimum. If $v''(t) < 0$, the point is a maximum.

First, let us solve $v'(t) = 0$. $v(t) = 5t^2 - 4t + 7$, so $v'(t) = 10t - 4$.

$$10t - 4 = 0$$
$$10t = 4$$
$$t = \frac{4}{10} = \frac{2}{5}$$

At $t = \frac{2}{5}$, the velocity is either at a minimum or maximum.

$v''(t) = 10$. Since $10 > 0$, the particle achieves a minimum velocity when $t = \frac{2}{5}$.

28. B

We know that velocity of an object is the derivative of its function position and that acceleration is the derivative of the velocity.

We are given that the acceleration due to gravity is -32 feet per second per second. Thus, the acceleration of the projectile is $a(t) = -32$. By integrating $a(t)$, we get the velocity of the object is $v(t) = -32t + 400$, since the initial velocity is 400 feet per second.

The projectile will reach its maximum height when $v(t) = 0$:

$$-32t + 400 = 0$$
$$-32t = -400$$
$$t = 12.5$$

So, at $t = 12.5$ seconds after initial firing, the projectile will reach its maximum height.

Next, we find the integral of $v(t) = -32t + 400$ to get the height of the projectile at $t = 12.5$ seconds after initial firing. We get $h(t) = -16t^2 + 400t + 100$ since the projectile started at a position 100 feet above ground level.

Thus, our answer is $h(12.5)$, which is 2600 feet.

29. A

We begin by labeling the figure supplied to us. Let the horizontal distance from the bottom of the wall to the bottom of the ladder be x and let the vertical distance from the top of the ladder to the ground be y. Then, by the Pythagorean Theorem, we know $x^2 + y^2 = 13^2$.

Since the bottom of the ladder is sliding at a rate of $\frac{1}{2}$ a foot per second, we also know that $\frac{dx}{dt} = \frac{1}{2}$. We want to find $\frac{dy}{dt}$, which is the rate at which the top of the ladder is sliding. We implicitly differentiate our equation to obtain $2x\frac{dx}{dt} + 2y\frac{dy}{dt} = 0.$

When $y = 5$, we know $x = 12$ (again thanks to the Pythagorean Theorem and the fact that the ladder is always 13 feet long).

Substituting all this information yields $2(12)(\frac{1}{2}) + 2(5)\frac{dy}{dt} = 0$, which gives answer (A) upon solving for $\frac{dy}{dt}$.

30. E

The position of the particle is given by

$$x(t) = \int v(t)\,dt = \int 5t^2 - 4t + 7\,dt = \frac{5}{3}t^3 - 2t^2 + 7t + C.$$

Solve for C by substituting $x(3) = 8$:

$$\frac{5}{3}(27) - 2(9) + 21 + C = 8$$
$$45 - 18 + 21 + C = 8$$
$$C = -40$$

Therefore, the position of the particle can be written

$$x(t) = \frac{5}{3}t^3 - 2t^2 + 7t - 40.$$

The total distance traveled is given by $\int_0^2 |v(t)|\,dt$. On this interval, though, $v(t)$ is positive, so we can simply write the integral as $\int_0^2 v(t)\,dt$ and solve:

$$\int_0^2 v(t)\,dt = x(t)\Big|_0^2 = \frac{5}{3}t^3 - 2t^2 + 7t - 40\Big|_0^2$$

$$= \left(\frac{5}{3}2^3 - 2\cdot 2^2 + 7\cdot 2 - 40\right) - \left(\frac{5}{3}0^3 - 2\cdot 0^2 + 7\cdot 0 - 40\right)$$

$$= \frac{40}{3} - 8 + 14$$

$$= \frac{58}{3}$$

DIAGNOSTIC CORRELATION CHART

Question No.	Topic	Chapter in Which the Topic Is Covered
1	Express Graph in Interval Notation	Ch. 1
2	Express Inequality in Interval Notation	Ch. 1
3	Simplifying Absolute Value Expression	Ch. 2
4	Solving Absolute Value Equation	Ch. 2
5	Graphing a Polynomial Function	Ch. 3
6	Solving a Polynomial Function	Ch. 3
7	Find the Slope of a Line	Ch. 4
8	Write the Equation of a Line	Ch. 4
9	Trigonometric Value of an Angle	Ch. 5
10	Solving a Trigonometric Equation	Ch. 5
11	Indeterminate Limits Using Algebraic Techniques	Ch. 6
12	Limits as x Approaches Infinity	Ch. 6
13	Limits of Trigonometric Functions	Ch. 6
14	Derivative of Polynomial Using Power Rule and Sum Rule	Ch. 7
15	Chain Rule and Trigonometric Functions	Ch. 7
16	Implicit Differentiation	Ch. 7

Question No.	Topic	Chapter in Which the Topic Is Covered
17	Writing Equation of a Tangent Line	Ch. 8
18	First Derivative and Increasing/Decreasing Behavior of f	Ch. 8
19	Second Derivative and Concavity	Ch. 8
20	Indefinite Integral of a Trigonometric Function	Ch. 9
21	Definite Integral of a Polynomial Function	Ch. 9
22	Fundamental Theorem of Calculus	Ch. 9
23	Area Between Two Curves	Ch. 10
24	Area Bounded By Curves	Ch. 10
25	Volume of Solids of Revolution	Ch. 10
26	Absolute Maximum and Absolute Minimum	Ch. 11
27	Velocity and Acceleration	Ch. 11
28	Optimization	Ch. 12
29	Related Rates	Ch. 12
30	Accumulated Change	Ch. 12

CHAPTER 1

Interval Notation

WHAT IS INTERVAL NOTATION?

An interval is the range of all real numbers in a set. Interval notation expresses this range using parentheses (( )) and/or brackets ([ ]) to enclose the endpoints of the range. A left parenthesis (( ) or a right parenthesis ( )) indicates that the endpoint is *not* included in the interval, while a left bracket ([ ) or a right bracket ( ]) indicates that the endpoint is included. For example, the interval "(4,9)" indicates the set of all real numbers *between* 4 and 9 but *not including* 4 and 9. Do not confuse this interval notation with the form (x,y) for the coordinates of a point on a Cartesian coordinate system.

CONCEPTS TO HELP YOU

1. Interval expressed as an inequality: We can express an interval as an inequality using the symbols $<$, $>$, $\leq$, and $\geq$. For example, the expression "$4 < x < 9$" means that the value of x is somewhere between 4 and 9, not including 4 and 9. The expression "$7 \leq x \leq 15$" means that the value of x is greater than or equal to 7 and less than or equal to 15. The numbers represent the endpoints of the intervals.

2. Interval represented on a number line: A range of real numbers can also be depicted graphically on a number line. An open circle (O) on the number line means that the corresponding number is not included in the interval. A solid circle (●) on the number line means that the corresponding number is included in the interval.

3. Types of intervals: The following table shows the relationship between interval notation and inequalities, as well as how they are graphed on a number line. Note that a and b are real numbers and $a < b$.

(a, b)	$a < x < b$	
$[a, b]$	$a \leq x \leq b$	
$[a, b)$	$a \leq x < b$	
$(a, b]$	$a < x \leq b$	
(a, ∞)	$x > a$	
$[a, \infty)$	$x \geq a$	
$(-\infty, b)$	$x < b$	
$(-\infty, b]$	$x \leq b$	

STEPS YOU NEED TO REMEMBER

1. Notation for the left endpoint.

If the left endpoint is included in the interval—that is, if $a \leq x$ or a is marked with a ● on the number line—then we write "$[a$".

If the left endpoint is *not* included in the interval—that is, if $a < x$ or a is marked with a O on the number line—then we write "$(a$".

If there is no left endpoint—that is, if $x < a$ or the graph extends without bound to the left—then we write "$(-\infty$".

2. Notation for the right endpoint.

If the right endpoint is included in the interval—that is, if $x \leq b$ or b is marked with a ● on the number line—then we write "$b]$".

If the right endpoint is *not* included in the interval—that is, if $x < b$ or b is marked with a O on the number line—then we write "$b)$".

If there is no right endpoint—that is, if $x > b$ or the graph extends without bound to the right—then we write "$\infty)$".

3. *Combine the notations.*

When you combine the notations for the left and right endpoints, you should get an interval that takes a form represented in the table in the previous page.

COMMON INTERVAL NOTATION QUESTIONS

Open Intervals: Solve: $6x + 5 > 3x - 1$. Express the answer in interval notation.

Step 1: Solve for x.

Step 2: Determine whether the endpoints are included in the interval. If you need to, visualize the inequality on a number line.

Step 3: Write the interval notation using a left parenthesis or a left bracket ((or [ ) for the left endpoint and right parenthesis or a right bracket ( ) or]) for the right endpoint. When writing the notation, you may find it helpful to refer to the table of types of interval notations.

Solution and Explanation: The solution to this problem is the interval $(-2,\infty)$.

When we solve for x, we get $x > -2$:

$$6x + 5 > 3x - 1$$
$$6x - 3x + 5 > 3x - 3x - 1$$
$$3x + 5 > -1$$
$$3x + 5 - 5 > -1 - 5$$
$$3x > -6$$
$$\frac{3x}{3} > \frac{-6}{3}$$
$$x > -2$$

On a number line, the interval has a O at −2 and extends to the right without bound.

−2

Because the endpoints are *not* included in the interval, we use left and right parentheses for the endpoints.

Note that we always accompany ∞ or −∞ with a *parenthesis*, never brackets. When an interval is enclosed by left *and* right parentheses, it is called an open interval.

The next problem involves another type of interval—the closed interval.

Closed Intervals: Solve: $-1 \leq 4x - 3$ and $4x - 3 \leq 9$.

Step 1: Combine the two inequalities and solve for x.

Step 2: Determine whether the endpoints are included in the interval. If you need to, visualize the inequality on a number line.

Step 3: Write the interval notation using a left parenthesis or a left bracket ((or [ ) for the left endpoint and right parenthesis or a right bracket ( ) or]) for the right endpoint. When writing the notation, you may find it helpful to refer to the table of types of interval notations.

Solution and Explanation: The solution to this problem is the interval $\left[\frac{1}{2}, 3\right]$.

We can combine the two inequalities because they both contain the expression $4x - 3$. Then we solve for x.

$$-1 \leq 4x - 3 \leq 9$$
$$-1 + 3 \leq 4x - 3 + 3 \leq 9 + 3$$
$$2 \leq 4x \leq 12$$
$$\frac{2}{4} \leq \frac{4x}{4} \leq \frac{12}{4}$$
$$\frac{1}{2} \leq x \leq 3$$

The number line has a ● at $\frac{1}{2}$ and at 3, meaning that both endpoints *are* included.

$$\frac{1}{2} \qquad\qquad 3$$

Therefore, we use "[" and "]" to enclose our interval. This type of interval is called a closed interval.

When one endpoint is included but the other is not, the interval is sometimes called a half-open or half-closed interval.

Half-Open Intervals: Solve: $2x + 10 > -12$ and $3x - 4 \le 8$. Express in interval notation.

Step 1: Solve for x in each equation, treating them separately.

Step 2: Find the intersection of the solution sets.

Step 3: Determine whether the endpoints are included in the interval. If you need to, visualize the inequality on a number line.

Step 4: Write the interval notation using a left parenthesis or a left bracket ((or [ ) for the left endpoint and right parenthesis or a right bracket ( ) or]) for the right endpoint. When writing the notation, you may find it helpful to refer to the table of types of interval notations.

Solution and Explanation: The solution to this problem is the interval $(-11, 4]$.

We solve for x in each equation.

$$
\begin{array}{ccc}
2x + 10 > -12 & & 3x - 4 \le 8 \\
2x + 10 - 10 > -12 - 10 & & 3x - 4 + 4 \le 8 + 4 \\
2x > -22 & \text{and} & 3x \le 12 \\
\dfrac{2x}{2} > \dfrac{-22}{2} & & \dfrac{3x}{3} \le \dfrac{12}{3} \\
x > -11 & & x \le 4
\end{array}
$$

When we visualize the graph of $x > -11$, we see that it begins at -11 and extends to the right. The graph of $x \le 4$ begins at 4 and extends to the left. The two inequalities intersect in the region between -11 and 4. In other words, $-11 < x \le 4$.

The interval includes the right endpoint but not the left endpoint, so we use (and] to enclose it. This is sometimes called a half-open interval.

Half-Closed Intervals: Solve: $3x - 2 < -11$ or $4x - 1 \geq 7$. Express in interval notation.

Step 1: Solve for each x separately.

Step 2: Find the union of the solution sets.

Step 3: Determine whether the endpoints are included in the interval. If you need to, visualize the inequality on a number line.

Step 4: Write the interval notation using a left parenthesis or a left bracket ((or [ ) for the left endpoint and right parenthesis or a right bracket ( ) or]) for the right endpoint. When writing the notation, you may find it helpful to refer to the table of types of interval notations.

Solution and Explanation: The solution to this problem is $(-\infty, -3) \cup [2, \infty)$.

When we solve for x, we get:

$$3x - 2 < -11 \qquad\qquad 4x - 1 \geq 7$$
$$3x - 2 + 2 < -11 + 2 \qquad 4x - 1 + 1 \geq 7 + 1$$
$$3x < -9 \qquad \text{or} \qquad 4x \geq 8$$
$$\frac{3x}{3} < \frac{-9}{3} \qquad\qquad \frac{4x}{4} \geq \frac{8}{4}$$
$$x < -3 \qquad\qquad x \geq 2$$

The graph of these two inequalities looks like this:

As you can see, the two inequalities do not intersect. One extends towards negative infinity $(-\infty, -3)$ and the other towards infinity $[2, \infty)$, so we cannot combine them into a single inequality. Instead, we use the symbol $\cup$ to denote all the possible values of x that belong to either of the two intervals.

Remember that we never use left or right brackets (i.e., never use [or]) when we are dealing with infinty. A *parenthesis* always accompanies ∞ or –∞.

The interval [2, ∞) is sometimes called a half-closed interval because it is enclosed in [and).

Now that you are familiar with the different types of interval notation, you can get more practice with them by taking the 10-item Chapter Quiz below. Afterward, check your answers and read the explanations.

CHAPTER QUIZ

For Questions 1–5, solve for x and express the solution in interval notation.

1. $x \geq 2.5$

   (A) $(-\infty, 2.5)$

   (B) $(-\infty, 2.5]$

   (C) $(2.5, \infty)$

   (D) $[2.5, \infty)$

   (E) $[2.5, \infty]$

2. $8x - 6 > 5x + 3$

   (A) $(-\infty, 3)$

   (B) $(3, \infty)$

   (C) $[-\infty, 3)$

   (D) $[3, \infty)$

   (E) $[3, \infty]$

3. $-4 \leq x + 6 \leq 11$

   (A) $(-10, 5)$

   (B) $(2, 17)$

   (C) $[-10, 5]$

   (D) $[-10, 5)$

   (E) $[2, 17]$

4. $x + 4 < 7$ or $2x \geq 10$

   (A) $(3, 5]$

   (B) $[5, 3)$

   (C) $(-\infty, 3) \cup (5, \infty)$

   (D) $(-\infty, 10] \cup (3, \infty)$

   (E) $(-\infty, 3) \cup [5, \infty)$

5. $16 < 4x - 4$ and $4(x - 1) \leq 20$

   (A) $(5, 6]$

   (B) $\left(\dfrac{17}{4}, \dfrac{21}{4} \right]$

   (C) $(5, 6)$

   (D) $(-\infty, 5) \cup [6, \infty)$

   (E) $(-\infty, 6] \cup (5, \infty)$

For Questions 6–9, express the interval notation as an inequality.

6. (–6,0]

(A) –6 < x < 0

(B) –6 ≤ x < 0

(C) –6 ≤ x ≤ 0

(D)   0 ≥ x > –6

(E)   0 > x ≥ –6

7. [–∞, 11]

(A) x < 11

(B) x > 11

(C) x ≥ 11

(D) x ≤ 11

(E) None of the above.

8. $\left[0, \dfrac{1}{2}\right)$

(A) $0 < x < \dfrac{1}{2}$

(B) $0 \le x < \dfrac{1}{2}$

(C) $0 \le x \le \dfrac{1}{2}$

(D) $\dfrac{1}{2} > x \ge 0$

(E) Both (B) and (D)

9. (–4,1)

(A) –4 < x < 1

(B)   1 < x < –4

(C) –4 < x and 1 > x

(D) –4 ≤ x ≤ 1

(E) Both (A) and (C)

10. Graph (–∞,–8) ∪ [8,∞)

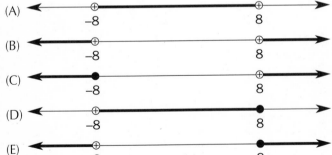

Answers and Explanations

1. D

The value of x begins at 2.5 and extends to the right without bound. The left endpoint is included in the interval because x can be equal to 2.5 and positive infinity is always accompanied by a). Therefore, we have the half-closed interval $[2.5, \infty)$.

2. B

When we solve for x, we get

$$8x - 6 > 5x + 3$$
$$8x - 5x - 6 + 6 > 5x - 5x + 3 + 6$$
$$3x > 9$$
$$x > 3$$

Here, the value of x extends to the right without bound but the left endpoint is not included. Therefore, we have the open interval $(3, \infty)$.

3. C

When we solve for x, we get

$$-4 \leq x + 6 \leq 11$$
$$-4 - 6 \leq x \leq 11 - 6$$
$$-10 \leq x \leq 5$$

The value of x lies between -10 and 5, including -10 and 5. Therefore, we have the closed interval $[-10, 5]$.

4. E

This problem involves a disjunction of inequalities. We cannot combine the two inequalities, so we must solve for x in each inequality, treating them separately.

$$
\begin{array}{ccc}
x + 4 < 7 & & 2x \geq 10 \\
x + 4 - 4 < 7 - 4 & \text{or} & \dfrac{2x}{2} \geq \dfrac{10}{2} \\
x < 3 & & x \geq 5
\end{array}
$$

We see that the left infinity begins at 3 but does not include 3, so we have an open interval ( ) with negative infinity. Negative infinity means that x is to the left of 3, not that it is necessarily a negative number. The right infinity begins at 5 and includes 5, so we have a half-closed interval [ ). The solution is $(-\infty,3) \cup [5, \infty)$ —the union of the two possible intervals for x.

5. A

Even though we have two inequalities, we see that $4x - 4 = 4(x - 1)$, so we can combine the two inequalities and solve for x.

$$16 < 4(x-1) \le 20$$
$$\frac{16}{4} < \frac{4(x-1)}{4} \le \frac{20}{4}$$
$$4+1 < x-1+1 \le 5+1$$
$$5 < x \le 6$$

We have a half-open interval here—(5,6]—because the left endpoint is not included, while the right endpoint is included in the interval.

6. D

The answer is $0 \ge x > -6$. We can rewrite this inequality as $-6 < x \le 0$, so that the smaller number is on the left. Now we can readily see that this inequality is equal to the interval notation (–6,0].

7. E

None of the answer choices are correct because the interval notation itself is written incorrectly. Remember that infinity and negative infinity are always accompanied by left or right parentheses ((or)). If the interval notation had been in the correct form, $(-\infty, 11]$, the correct inequality would have been $x \le 11$, choice (D).

8. E

The left bracket tells us that 0 is included interval and the right parenthesis tells us that $\frac{1}{2}$ is not included. Choice (B)— $0 \le x < \frac{1}{2}$ — expresses this interval correctly. However, the method of rewriting the inequality that we used in Question 6 means that choice (D) — $\frac{1}{2} > x \geq 0$—is the same as (B). Thus both (B) and (D) are correct, which means (E) is the correct answer.

9. E

When we combine the inequalities in choice (C)— $-4 < x$ and $1 > x$—we get the inequality in choice (A): $-4 < x < 1$. Both choices express the interval between -4 and 1 but do not include -4 or 1. This means that (E) is the correct answer.

10. E

The two intervals extend in opposite directions, one towards negative infinity and the other towards positive infinity, so the graph cannot correspond to choice (A) or choice (D).

We know that -8 is not included because it is accompanied by a right parenthesis, so there should be a O on the -8 on the number line. We know that 8 is included because it is accompanied by a left bracket, so there should be a ● on the 8 on the number line.

Absolute Value

WHAT IS ABSOLUTE VALUE?

The absolute value of a number is the distance of that number from 0 on the number line. For example, −3 is 3 units from 0 and +3 is also 3 units from 0.

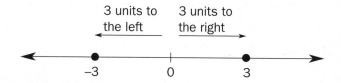

3 units to
the left

3 units to
the right

−3   0   3

The absolute value is denoted by the symbols $|$ and $|$.

CONCEPTS TO HELP YOU

1. $|a|$ is always positive: Distance is always positive because an object cannot be a negative distance from something. Therefore, the absolute value of any real number a is always positive. For example, $|-4| = 4$ and $|4| = 4$.

2. Properties of absolute value: For any real numbers a and b:

    $$|ab| = |a| \times |b|$$

    $$\left|\frac{a}{b}\right| = \frac{|a|}{|b|}$$

    $$|-a| = |a|$$

3. For every $|a|$, there are two possible values for a:

    $$|a| = \begin{cases} a \text{ if } a \geq 0 \\ -a \text{ if } a < 0 \end{cases}$$

    The negative sign in $-a$ indicates the *opposite* of a for $a < 0$.

STEPS YOU NEED TO REMEMBER

1. *Remove the absolute value symbols.*

To solve absolute value equations and inequalities such as $|x + a| = b$ or $|x + a| < b$, we first need to remove the absolute value symbols $|$ and $|$.

2. *Set up two equations or inequalities.*

Recall that for every $|a|$, there are two possible values for a.

If $|ax| = b$, then $ax = b$ and $ax = -b$.

If $|x \pm a| = b$, then $x \pm a = b$ and $x \pm a = -b$.

If $|x \pm a| < b$, then $x \pm a < b$ and $x \pm a > -b$.

If $|x \pm a| > b$, then $x \pm a > b$ or $x \pm a < -b$.

If $|x \pm a| \leq b$, then $x \pm a \leq b$ and $x \pm a \geq -b$.

If $|x \pm a| \geq b$, then $x \pm a \geq b$ or $x \pm a \leq -b$.

Note that the negative sign in $-b$ above does not necessarily denote a negative number. Rather, it indicates a value *opposite* to that of b.

3. *Solve for x.*

Once we have set up the appropriate equations or inequalities, we can solve for x algebraically.

If we are working with equations, the solution set will consist of two possible values for x.

If we are working with inequalities, the solution will be an interval.

COMMON ABSOLUTE VALUE QUESTIONS

Absolute Value of a Single Term: Simplify $|-6x|$.

Step 1: Make the negative number positive by removing the negative sign.

Step 2: Remove the absolute value symbols $|$ and $|$.

> **Solution and Explanation:** We know that absolute value is always positive. We also know that $|-a| = |a|$ (property of absolute value). Therefore, $|-6x| = |6x| = 6x$.

Because we only have an algebraic expression here, we cannot solve for x. However, we can substitute values for x to see that $|-6x| = 6x$.

Suppose $x = 2$.

Then $|-6 \times 2| = |-12| = |12| = 12$.

If $x = -2$, then $|-6 \times -2| = |12| = 12$.

The next problem involves an equation in which we can solve for x.

Equations with One Absolute Value Expression:
Solve: $\left| x - 7 \right| = 10$.

Step 1: Remove the absolute value symbols $|$ and $|$.

Step 2: Set up two equations in the forms $x + a = b$ and $x + a = -b$.

Step 3: Solve for x.

Solution and Explanation: Based on the definition of absolute value, we know that $x - 7$ is 10 units away from 0. What we don't know is whether $x - 7$ is to the *left* or the *right* of 0. In other words, we don't know if $x - 7 = 10$ or $x - 7 = -10$.

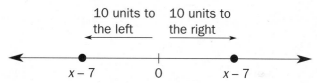

To find x, we need to solve both equations.

$$x - 7 = 10 \qquad\qquad x - 7 = -10$$

$$x - 7 + 7 = 10 + 7 \text{ and } x - 7 + 7 = -10 + 7$$

$$x = 17 \qquad\qquad x = -3$$

The solution is $\{-3, 17\}$.

Let's make sure our solution set satisfies the absolute value equation:

$$|x - 7| = 10 \qquad |x - 7| = 10$$

$$|-3 - 7| = 10 \quad\text{and}\quad |17 - 7| = 10$$

$$|-10| = 10 \qquad |10| = 10$$

$$10 = 10 \qquad\qquad 10 = 10$$

What if you have an equation involving *two* absolute value expressions?

Equations with Two Absolute Value Equations:
Solve: $\left|3x - 2\right| = \left|x + 5\right|$.

Step 1: Remove the absolute value symbols $|$ and $|$.

Step 2: Set up two equations in the forms $x + a = b$ and $x + a = -b$.

Step 3: Solve for x.

> **Solution and Explanation:** Notice that the steps involved in solving this problem are the same as the ones used in the previous question. Do not be confused by the double set of absolute value symbols. The concept remains the same. The point denoted by $3x - 2$ is at a distance of $x + 5$ units to the *left* or the *right* of 0.

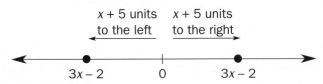

From the graph, we can see that

$$3x - 2 = x + 5$$
$$3x - x - 2 + 2 = x - x + 5 + 2$$
$$2x = 7$$
$$x = \frac{7}{2}$$

$$3x - 2 = -(x + 5)$$
$$3x - 2 = -x - 5$$
and $3x + x - 2 + 2 = -x + x - 5 + 2$
$$4x = -3$$
$$x = \frac{-3}{4}$$

The solution set is $\left\{\dfrac{-3}{4}, \dfrac{7}{2}\right\}$.

Double-check:

$$\left|3x - 2\right| = \left|x + 5\right|$$
$$\left|3\left(\frac{-3}{4}\right) - 2\right| = \left|\frac{-3}{4} + 5\right|$$
$$\left|\frac{-9}{4} - \left(\frac{2 \times 4}{4}\right)\right| = \left|\frac{-3}{4} + \left(\frac{5 \times 4}{4}\right)\right| \quad \text{and}$$
$$\left|\frac{-17}{4}\right| = \left|\frac{17}{4}\right|$$
$$\frac{17}{4} = \frac{17}{4}$$

$$\left|3x - 2\right| = \left|x + 5\right|$$
$$\left|3\left(\frac{7}{2}\right) - 2\right| = \left|\frac{7}{2} + 5\right|$$
$$\left|\frac{21}{2} - \left(\frac{2 \times 2}{2}\right)\right| = \left|\frac{7}{2} + \left(\frac{5 \times 2}{2}\right)\right|$$
$$\left|\frac{17}{2}\right| = \left|\frac{17}{2}\right|$$
$$\frac{17}{2} = \frac{17}{2}$$

Now let's move on to absolute value inequalities.

Absolute Value Inequalities: Solve: $\left|5x + 3\right| < 18$. Write the answer in interval notation.

Step 1: Remove the absolute value symbols $|$ and $|$.

Step 2: Set up two inequalities in the forms $x + a < b$ and $x + a > -b$.

Step 3: Solve for x.

> **Solution and Explanation:** $5x + 3$ is a point on the number line that is less than 18 units from 0 but we don't know if it is to the *left* or the *right* of 0. Therefore, we need to solve for both possibilities. Let's visualize $5x + 3$ as two points on a number line.

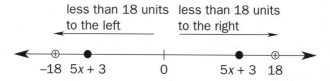

From the graph, we see that $5x + 3 > -18$ and $5x + 3 < 18$. We can combine the two equations and solve for $-18 < 5x + 3 < 18$.

$$-18 < 5x + 3 < 18$$
$$-18 - 3 < 5x < 18 - 3$$
$$-21 < 5x < 15$$
$$\frac{-21}{5} < x < 3$$

The solution in interval notation is $\left(\dfrac{-21}{5}, 3\right)$.

Double-check the solution by choosing a value for x that lies somewhere on the interval, such as -2:

$$\left|5x + 3\right| < 18$$
$$\left|5(-2) + 3\right| < 18$$
$$\left|-10 + 3\right| < 18$$
$$\left|-7\right| < 18$$
$$7 < 18$$

Thus far, we have used absolute value to denote the distance of a value from 0 on the number line. Absolute value can also denote distance between any two points on a number line.

Distance between Two Points: Find the distance between –5 and 13 on the number line.

Step 1: Subtract the two numbers within the absolute value symbols.

Step 2: If the difference of the two numbers is negative, make it positive.

Step 3: Drop the absolute value symbols.

> **Solution and Explanation:** It does not matter which number we subtract from the other; the solution remains the same.
>
> $|-5 - 13| = |-18| = |18| = 18$
>
> $|13 - (-5)| = |13 + 5| = |18| = 18$
>
> Understanding the properties of absolute value and knowing how to set up absolute value equations and inequalities will enable to you solve the problems in the Chapter Quiz below.

CHAPTER QUIZ

1. Find the distance between $-4x$ and $-7x$ on the number line.

   (A) $\dfrac{36}{5}$

   (B) $\dfrac{|36|}{|5|}$

   (C) $\left|\dfrac{-36}{5}\right|$

   (D) 7.2

   (E) All of the above.

2. Solve $|-5x| = 30$

   (A) $\{-6\}$

   (B) $\{6\}$

   (C) $(-6,6)$

   (D) $[-6,6]$

   (E) Both (A) and (B).

3. Solve $|4x - 2| = 6$

   (A) $\{-2,1\}$

   (B) $\{-1,2\}$

   (C) $\{-2,-1\}$

   (D) $\{1,2\}$

   (E) None of the above.

4. Solve $\left|x + \dfrac{1}{3}\right| = -3$

   (A) $\left\{\dfrac{-10}{3}, \dfrac{8}{3}\right\}$

   (B) $\left\{\dfrac{-8}{3}, \dfrac{10}{3}\right\}$

   (C) $\left\{\dfrac{-10}{3}, \dfrac{-8}{3}\right\}$

   (D) $\left\{\dfrac{8}{3}, \dfrac{10}{3}\right\}$

   (E) None of the above.

5. Solve $|2x + 1| > 7$

  (A) $\{-4,3\}$

  (B) $(-4,\infty)$

  (C) $(-\infty,-4) \cup (3, \infty)$

  (D) $(-\infty,-4) \cap (3, \infty)$

  (E) None of the above.

6. Solve $|2x + 4| \le 2$

  (A) $[-3, \infty)$

  (B) $(-1,\infty)$

  (C) $(-\infty,-3) \cup (-1, \infty)$

  (D) $[-3,-1]$

  (E) None of the above.

7. Solve $|5x - 7| = |x + 10|$

  (A) $\left\{\dfrac{-1}{2}, \dfrac{17}{4}\right\}$

  (B) $\left\{\dfrac{17}{6}, \dfrac{17}{4}\right\}$

  (C) $\left\{\dfrac{17}{4}\right\}$

  (D) $\left\{\dfrac{1}{2}, \dfrac{17}{4}\right\}$

  (E) $\left\{\dfrac{-17}{4}, \dfrac{1}{2}\right\}$

8. Solve $|6x + 5| = |5x + 11|$

  (A) $\left\{\dfrac{-16}{11}, 6\right\}$

  (B) $\left\{-6, \dfrac{16}{11}\right\}$

  (C) $\left\{\dfrac{16}{11}, 6\right\}$

  (D) $\left\{-6, \dfrac{-16}{11}\right\}$

  (E) $\{6\}$

9. Solve $|-5x| < 25$

  (A) $(-\infty,-5) \cup (5, \infty)$

  (B) $(-5,5)$

  (C) $x < -5$ and $x < 5$

  (D) $x > -5$ and $x > 5$

  (E) None of the above.

10. Solve $|-7x + 4| \ge 18$

  (A) $(-\infty,-2] \cup \left(\dfrac{22}{7},\infty\right)$

  (B) $(-\infty,-2] \cap \left(\dfrac{22}{7},\infty\right)$

  (C) $(-\infty,-2] \cup \left[\dfrac{22}{7},\infty\right)$

  (D) $\left(-2,\dfrac{22}{7}\right)$

  (E) None of the above.

Answers and Explanations

1. E

$$\left|-4x - (-7x)\right| = \left|-4x + 7x\right| = \left|3x\right| = 3x$$

Alternatively, $\left|-7x - (-4x)\right| = \left|-7x + 4x\right| = \left|-3x\right| = 3x.$

Let's double-check. Suppose $x = 3$. Then $3x = 9$ and

$$\left|-4(3) - (-7(3))\right| = \left|-12 + 21\right| = \left|9\right| = 9.$$

2. E

If $\left|ax\right| = b$, then $ax = b$ and $ax = -b$. Therefore,

$$-5x = 30 \quad \text{and} \quad -5x = -30$$
$$x = -6 \qquad\qquad x = 6$$

Choices (A) and (B) each contain only half of the solution set. The solution set is $\{-6,6\}$, which is a combination of choices (A) and (B).

3. B

If $\left|x + a\right| = b$, then $x + a = b$ and $x + a = -b$. Therefore,

$$4x - 2 = 6 \qquad 4x - 2 = -6$$
$$4x = 8 \quad \text{and} \quad 4x = -4$$
$$x = 2 \qquad\qquad x = -1$$

The solution set is $\{-1,2\}$.

Double-check:

$$\left|4x - 2\right| = 6 \qquad\qquad \left|4x - 2\right| = 6$$
$$\left|4(-1) - 2\right| = 6 \qquad \left|4(2) - 2\right| = 6$$
$$\left|-4 - 2\right| = 6 \quad \text{and} \quad \left|8 - 2\right| = 6$$
$$\left|-6\right| = 6 \qquad\qquad \left|6\right| = 6$$
$$6 = 6 \qquad\qquad\qquad 6 = 6$$

4. E

We cannot solve this problem because absolute value can never be equal to a negative number: $\left| x + \frac{1}{3} \right| \neq -3$.

5. C

If $|x \pm a| > b$, then $x \pm a > b$ or $x \pm a < -b$. Therefore,

$$2x + 1 > 7 \qquad 2x + 1 < -7$$
$$2x > 6 \quad \text{or} \quad 2x < -8$$
$$x > 3 \qquad\qquad x < -4$$

The graph of these two inequalities looks like this:

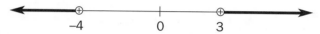

Recall from Chapter 1 that the interval notation for $x > 3$ is $(3, \infty)$ and the interval notation for $x < -4$ is $(-\infty, -4)$. Also, the solution set for a disjunction of inequalities is the union of the individual solution sets. Hence, our solution for this problem in interval notation is $(-\infty, -4) \cup (3, \infty)$.

We can double-check our answer by choosing a number greater than 3 (such as 5) and a number less than –4 (such as –6) and substituting these values into our original equation:

$$|2x + 1| > 7 \qquad\qquad |2x + 1| > 7$$
$$|2(5) + 1| > 7 \qquad\quad |2(-6) + 1| > 7$$
$$|10 + 1| > 7 \quad \text{or} \quad |-12 + 1| > 7$$
$$|11| > 7 \qquad\qquad\quad |-11| > 7$$
$$11 > 7 \qquad\qquad\quad 11 > 7$$

The $\cap$ symbol in choice (D) means intersection, which does not apply here because $x > 3$ and $x < -4$ do not intersect.

6. D

If $|x \pm a| \leq b$, then $x \pm a \leq b$ and $x \pm a \geq -b$. Therefore,

$$2x + 4 \leq 2 \qquad\qquad 2x + 4 \geq -2$$
$$2x \leq -2 \quad \text{and} \quad 2x \geq -6$$
$$x \leq -1 \qquad\qquad x \geq -3$$

The graph of these two inequalities looks like this:

We can see that the two inequalities intersect, so the value of x is between -3 and -1 and includes both endpoints.

We can double-check by substituting $x = -2$ into our original equation:

$$|2x + 4| \leq 2$$
$$|2(-2) + 4| \leq 2$$
$$|-4 + 4| \leq 2$$
$$|0| \leq 2$$
$$0 \leq 2$$

7. A

We solve this equation the same way we solved Question 3.

$$|5x - 7| = |x + 10|$$

$$|5x - 7| = |x + 10| \qquad\qquad 5x - 7 = -(x + 10)$$
$$5x - 7 = x + 10 \qquad\qquad 5x - 7 = -x - 10$$
$$5x = x + 17 \quad \text{and} \quad 5x = -x - 3$$
$$4x = 17 \qquad\qquad 6x = -3$$
$$x = \frac{17}{4} \qquad\qquad x = \frac{-3}{6} = \frac{-1}{2}$$

The solution set is $\left(\dfrac{-1}{2}, \dfrac{17}{4}\right)$.

Double-check:

$$\left|5\left(\frac{17}{4}\right)-7\right|=\left|\frac{17}{4}+10\right|$$
$$\left|\frac{85}{4}-\left(\frac{17\times4}{4}\right)\right|=\left|\frac{17}{4}+\left(\frac{10\times4}{4}\right)\right|$$
$$\left|\frac{57}{4}\right|=\left|\frac{57}{4}\right|$$
$$\frac{57}{4}=\frac{57}{4}$$

and

$$\left|5\left(\frac{-1}{2}\right)-7\right|=\left|\frac{-1}{2}+10\right|$$
$$\left|\frac{-5}{2}-\left(\frac{7\times2}{2}\right)\right|=\left|\frac{-1}{2}+\left(\frac{10\times2}{2}\right)\right|$$
$$\left|\frac{-19}{2}\right|=\left|\frac{19}{2}\right|$$
$$\frac{19}{2}=\frac{19}{2}$$

8. A

Set up the two equations and solve for x:

$$6x+5=-(5x+11)$$

$$6x+5=5x+11 \qquad 6x+5=-5x-11$$
$$6x=5x+6 \quad \text{and} \quad 6x=-5x-16$$
$$x=6 \qquad\qquad 11x=-16$$
$$x=\frac{-16}{11}$$

The solution set is $\left\{\dfrac{-16}{11},6\right\}$.

Double-check the solutions:

$$\left|6(6)+5\right|=\left|5(6)+11\right|$$
$$\left|36+5\right|=\left|30+11\right|$$
$$\left|41\right|=\left|41\right|$$
$$41=41$$

and

$$\left|6\left(\frac{-16}{11}\right)+5\right|=\left|5\left(\frac{-16}{11}\right)+11\right|$$
$$\left|\left(\frac{-96}{11}\right)+\left(\frac{5\times11}{11}\right)\right|=\left|\left(\frac{-80}{11}\right)+\left(\frac{11\times11}{11}\right)\right|$$
$$\left|\frac{-41}{11}\right|=\left|\frac{41}{11}\right|$$
$$\frac{41}{11}=\frac{41}{11}$$

9. B

We set up our two equations and solve for x. However, whenever you divide or multiply both sides of an inequality by a negative number, you need to reverse the direction of the inequality sign.

$$-5x < 25 \qquad -5x > -25$$

$$\frac{-5x}{-5} > \frac{25}{-5} \quad \text{and} \quad \frac{-5x}{-5} < \frac{-25}{-5}$$

$$x > -5 \qquad x < 5$$

The two inequalities intersect and the interval is $(-5,5)$. We can double-check by choosing a value for x that lies on this interval, such as -3.

$$\left|-5(-3)\right| < 25$$

$$\left|-15\right| < 25$$

$$15 < 25$$

10. C

When we solve for x, we get

$$-7x + 4 \geq 18 \qquad -7x + 4 \leq -18$$

$$-7x \geq 14 \quad \text{or} \quad -7x \leq -22$$

$$x \leq -2 \qquad x \geq \frac{22}{7}$$

These two inequalities do not intersect, as you can see on the graph below:

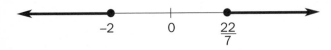

Also, one interval is half-open and the other is half-closed. The correct solution is the union of the two intervals: $(-\infty, -2] \cup \left[\frac{22}{7}, \infty\right)$.

Double-check using $x = -3$ and $x = 4$:

$$\left|-7(-3)+4\right| \geq 18 \qquad \left|-7(4)+4\right| \geq 18$$

$$\left|21+4\right| \geq 18 \qquad \left|-28+4\right| \geq 18$$

$$\left|25\right| \geq 18 \quad \text{or} \quad \left|-24\right| \geq 18$$

$$25 \geq 18 \qquad 24 \geq 18$$

Functions

WHAT ARE FUNCTIONS?

A function is a relationship between two quantities (x and y) such that, for every value of x, there is one and only one value for y. We can use the vertical line test to determine whether or not a relation is a function.

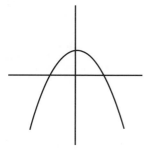

Passes vertical line test

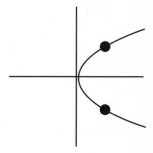

Fails vertical line test

The graph on the left is a function because the vertical line intersects the curve at only one point, indicating that each value of x has only one corresponding value of y. However, the graph on the right cannot be a function because for each value of x, the vertical line passes through a positive and negative y.

CONCEPTS TO HELP YOU

1. Polynomial functions: A polynomial function consists of a polynomial expression in x set equal to y or $f(x)$, read as "f of x." For example, $y = 2x^2 + 4x - 7$ can also be written as $f(x) = 2x^2 + 4x - 7$.

2. Types of polynomial functions:

   A polynomial function of degree zero is called a constant function ($y = a$).

   A polynomial function of degree one is called a linear function ($y = x$).

   A polynomial function of degree two is called a quadratic function ($y = x^2$).

   A polynomial function of degree three is called a cubic function ($y = x^3$).

3. Graphs of polynomial functions: If you know the values of x and their corresponding values of $f(x)$, you can plot the ordered pairs ($x, f(x)$) on an xy coordinate plane.

STEPS YOU NEED TO REMEMBER

Evaluating a function for a given value of x

1. Substitute the value of x into the polynomial expression.

To solve for $f(x)$ when you are given the value of x, first substitute the value of x into the polynomial. For example, for $f(x) = 3x^3 - 4$ and $x = 2$, $f(2) = 3(2)^3 - 4$.

2. Carry out the calculations.

Once you have substituted the value of x into the polynomial, you can carry out the calculations to get the value of $f(x)$. Using the example above, $f(2) = 3(2)^3 - 4 = 20$.

Solving for x

1. Set the polynomial equal to zero.

When solving for x of a polynomial function, set the polynomial equal to zero. For example, to solve for $f(x) = x^2 - 8x + 12$, we write $x^2 - 8x + 12 = 0$.

2. Factor the polynomial.

To factor the polynomial, we find two integers whose product equals 9 and whose sum equals –6. Through trial and error, we find that those two integers are –6 and –2. Therefore, the factors of $x^2 - 8 + 12$ are ($x - 6$) and ($x - 2$), and ($x - 6$)($x - 2$) = 0.

3. Set each factor equal to zero.

The Principle of Zero Products states that if a and b are real numbers and $ab = 0$, then either $a = 0$ or $b = 0$. Therefore, at least one of the factors of the polynomial must be equal to zero in order for the polynomial to be equal to zero. In other words, $x - 6 = 0$ or $x - 2 = 0$.

4. Solve for x.

Once we have set the factors equal to zero, we can solve for x. If $x - 6 = 0$, then $x = 6$. If $x - 2 = 0$, then $x = 2$. Therefore, the solution for the function $f(x) = x^2 - 8x + 12$ is $x = 6$ or $x = 2$.

COMMON FUNCTION QUESTIONS

Graphing Common Functions: Graph the following functions:

$$f(x) = x$$
$$f(x) = x^2$$
$$f(x) = x^3$$
$$f(x) = |x|$$
$$f(x) = \sqrt{x}$$
$$f(x) = \frac{1}{x}$$

Step 1: Make a table for the values for x and $f(x)$.

Step 2: Choose values for x and plug into the polynomial expression.

Step 3: Compute for $f(x)$.

Step 4: Plot the resulting coordinate points $(x, f(x))$ on an xy plane.

Step 5: Connect the points.

> **Solution and Explanation:** To graph each function, you need to create a table of values for x and $f(x)$. First choose values for x that are simple to work with, such as -2, -1, 0, 1, and 2. When you plug the values of x into the polynomial and perform the operations of the expression, you get the values of $f(x)$. Once you have some ordered pairs $(x, f(x))$, you can graph the function. Be sure to plot as many points as you need to get a clear picture of the graph.

$f(x) = x$

In this function, the values of $f(x)$ and x are exactly the same.

x	$f(x)$	$(x,f(x))$
–2	–2	(–2,–2)
–1	–1	(–1,–1)
0	0	(0,0)
1	1	(1,1)
2	2	(2,2)

Our $(x,f(x))$ coordinate pairs are (–2,–2), (–1,–1), (0,0), (1,1), and (2,2). We plot these coordinates on a coordinate system and connect the points.

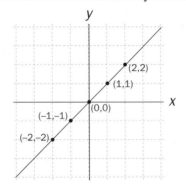

We see that $f(x) = x$ is a straight line. In fact, *all* graphs of polynomial functions of degree one are straight lines. We will investigate linear functions in the next chapter.

$f(x) = x^2$

We have a quadratic function here because the polynomial is of degree two.

Table of values:

x	$f(x)$	$(x,f(x))$
–2	$(-2)^2 = 4$	(–2,4)
–1	$(-1)^2 = 1$	(–1,1)
0	$(0)^2 = 0$	(0,0)
1	$(1)^2 = 1$	(1,1)
2	$(2)^2 = 4$	(2,4)

Now that we have our coordinates, we can graph $f(x) = x^2$.

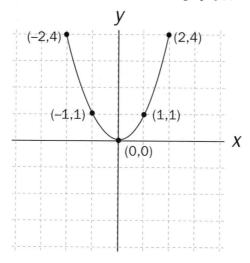

This U-shaped curve is called a parabola. The graph of any quadratic function is a parabola.

$f(x) = x^3$

A polynomial of degree three is called a cubic function and $f(x) = x^3$ is the most basic cubic function.

Table of values:

x	$f(x)$	$(x, f(x))$
-2	$(-2)^3 = -8$	$(-2, -8)$
-1	$(-1)^3 = -1$	$(-1, -1)$
0	0	$(0, 0)$
1	$(1)^3 = 1$	$(1, 1)$
2	$(2)^3 = 8$	$(2, 8)$

Graph of function:

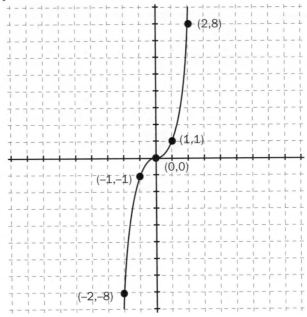

Graphs of any cubic function will be a variation of the above graph.

$f(x) = |x|$

Recall from the previous chapter that the absolute value of a number is always positive. Therefore, $f(x)$ is always positive.

Table of values:

x	$f(x)$	$(x,f(x))$
-2	$|-2| = 2$	$(-2,2)$
-1	$|-1| = 1$	$(-1,1)$
0	0	$(0,0)$
1	$|1| = 1$	$(1,1)$
2	$|2| = 2$	$(2,2)$

Graph of function:

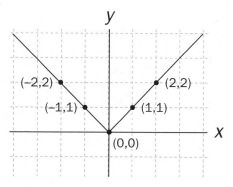

$f(x) = \sqrt{x}$

For the square root function, we will choose values for x that will allow us to easily compute $\sqrt{x}$.

Table of values:

x	$f(x)$	$(x, f(x))$
0	0	(0,0)
1	$\sqrt{1} = 1$	(1,1)
4	$\sqrt{4} = 2$	(4,2)

Graph of function:

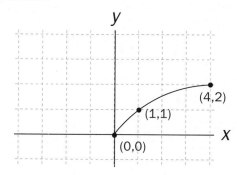

$f(x) = \dfrac{1}{x}$

Table of values:

x	$f(x)$	$(x, f(x))$
-2	$\dfrac{1}{-2}$	$\left(-2, \dfrac{1}{-2}\right)$
-1	$\dfrac{1}{-1} = -1$	$(-1, -1)$
$\dfrac{-1}{2}$	$\dfrac{1}{\frac{-1}{2}} = -2$	$\left(\dfrac{-1}{2}, -2\right)$
0	$\dfrac{1}{0} =$ undefined	does not exist
$\dfrac{1}{2}$	$\dfrac{1}{\frac{1}{2}} = 2$	$\left(\dfrac{1}{2}, 2\right)$
1	$\dfrac{1}{1} = 1$	$(1, 1)$
2	$\dfrac{1}{2}$	$\left(2, \dfrac{1}{2}\right)$

Notice that $f(x)$ is undefined when $x = 0$ because then the denominator is zero. Also, if you plotted enough points, you will see that $f(x) = \dfrac{1}{x}$ gets closer and closer to zero but never reaches it.

Graph of function:

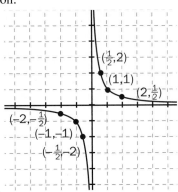

You have just graphed 6 types of functions by putting in x values and evaluating for $f(x)$. It is very important that you memorize these basic graphs because you will encounter variations of them frequently in calculus courses and throughout this book.

For a graph that crosses the x-axis, the value of $f(x)$ is 0 at the point where the graph intercepts the x-axis. The x-value of that point is called the zero of $f(x)$.

In this next problem, you will learn how to find zeros of a function—in other words, you'll learn how to solve for x when $f(x) = 0$.

Zeros of Functions: Find the zeros of the following functions:

$$f(x) = x^2 + 4x + 3$$
$$f(x) = x^3 - x^2 - 2x$$
$$f(x) = 2x^2 - 3x - 10$$

Step 1: Set the equation equal to zero.

Step 2: Factor the polynomial expression.

Step 3: Use the Principle of Zero Products.

Step 4: Solve for x.

Solution and Explanation:

$$f(x) = x^2 + 4x + 3$$

When we set the function equal to zero, we get the equation $x^2 + 4x + 3 = 0$. To factor the polynomial expression, we find two integers whose product equals 3 and whose sum equals 4. Those two integers are 3 and 1.

$$x^2 + 4x + 3 = 0$$
$$(x + 3)(x + 1) = 0$$

For the equation to be equal to zero, at least one of the binomials must equal zero. This follows from the Principle of Zero Products, which states that if a and b are real numbers and $ab = 0$, then either $a = 0$ or $b = 0$.

Therefore, we set each binomial equal to zero and solve for x.

$$x + 3 = 0 \quad \text{or} \quad x + 1 = 0$$
$$x = -3 \qquad x = -1$$

The solutions are $x = -3$ and $x = -1$.

Check to see that –3 and –1 both give a zero value for $f(x)$.

$$f(-3) = (-3)^2 + 4(-3) + 3 = 9 - 12 + 3 = 0$$
$$f(-1) = (-1)^2 + 4(-1) + 3 = 1 - 4 + 3 = 0$$

Since $f(x) = x^2 + 4x + 3$ is a quadratic function, its graph is a parabola that intercepts the x-axis at $(-3,0)$ or $(-1,0)$.

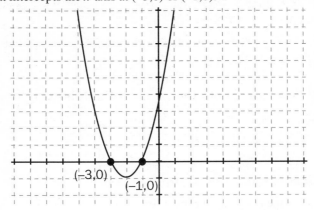

$$f(x) = x^3 - x^2 - 2x$$

Set the function equal to zero: $x^3 - x^2 - 2x = 0$.

Next, factor out the x in the cubic expression and then factor the resulting quadratic expression:

$$x^3 - x^2 - 2x = 0$$
$$x\left(x^2 - x - 2\right) = 0$$
$$x(x+1)(x-2) = 0$$

Using the Principle of Zero Products, we set the monomial and binomials equal to zero and solve for x:

$$x = 0 \text{ or } \begin{array}{c} x+1=0 \\ x=-1 \end{array} \text{ or } \begin{array}{c} x-2=0 \\ x=2 \end{array}$$

The solution is $x = 0$, $x = -1$, or $x = 2$.

Check to see if each of the values of x gives a zero value for $f(x)$.

$$f(0) = (0)^3 - (0)^2 - 2(0) = 0$$

$$f(-1) = (-1)^3 - (-1)^2 - 2(-1) = -1 - 1 + 2 = 0$$

$$f(2) = (2)^3 - (2)^2 - 2(2) = 8 - 4 - 4 = 0$$

When we graph the function, we see that the curve does in fact intercept the x-axis at $x = 0, -1$, and 2.

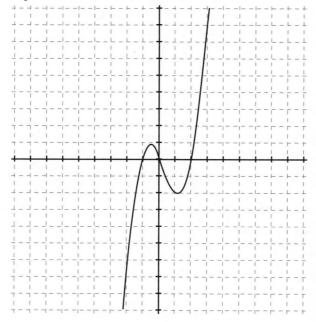

Sometimes, you will need to solve for functions in which the quadratic equation cannot be easily factored. You will need to use the quadratic formula to solve for x. For any quadratic equation in standard form (that is, in the form $ax^2 + bx + c = 0$),

$$x = \frac{-b \pm \sqrt{b^2 - 4ac}}{2a}.$$

$$f(x) = 2x^2 - 3x - 10$$

$$2x^2 - 3x - 10 = 0$$

Because we cannot factor $2x^2 - 3x - 10$, we need to use the quadratic formula.

Substitute 2 for a, -3 for b, and -10 for c.

$$x = \frac{-b \pm \sqrt{b^2 - 4ac}}{2a}$$

$$= \frac{-(-3) \pm \sqrt{(-3)^2 - 4(2)(-10)}}{2(2)}$$

$$= \frac{3 \pm \sqrt{89}}{4}$$

The zeros of $f(x) = 2x^2 - 3x - 10$ are $\dfrac{3 + \sqrt{89}}{4}$ and $\dfrac{3 - \sqrt{89}}{4}$.

You now know how to solve for x of a polynomial function $f(x)$. Now you will learn how to combine two different polynomial functions $f(x)$ and $g(x)$.

Operations with Functions: For $f(x) = 2x^2 + 5x - 12$ and $g(x) = 3x + 12$, find:

$(f + g)(x)$

$(f - g)(x)$

$(f \bullet g)(x)$

$\left(\dfrac{f}{g} \right)(x)$

Step 1: Set $(f + g)(x) = f(x) + g(x)$

Set $(f - g)(x) = f(x) - g(x)$

Set $(f \bullet g)(x) = f(x) \bullet g(x)$

Set $\left(\dfrac{f}{g} \right)(x) = \dfrac{f(x)}{g(x)}$

Step 2: Simplify.

Solution and Explanation:

$(f + g)(x)$

$$(f+g)(x) = f(x) + g(x)$$
$$= \left(2x^2 + 5x - 12\right) + \left(3x + 12\right)$$
$$= 2x^2 + 5x + 3x - 12 + 12$$
$$= 2x^2 + 8x$$

We will choose a value for x to check our answer.

$$f(2) = 2(2)^2 + 5(2) - 12 = 8 + 10 - 12 = 6$$
$$g(2) = 3(2) + 12 = 18$$
$$6 + 8 = 24$$
$$(f + g)(2) = 2(2)^2 + 8(2) = 8 + 16 = 24$$

$(f - g)(x)$

$$(f-g)(x) = f(x) - g(x)$$
$$= \left(2x^2 + 5x - 12\right) - \left(3x + 12\right)$$
$$= 2x^2 + 5x - 3x - 12 - 12$$
$$= 2x^2 + 2x - 24$$

Double-check:

$$f(2) = 6 \text{ and } g(2) = 18$$
$$6 - 18 = -12$$
$$(f + g)(2) = 2(2)^2 + 2(2) - 24 = 8 + 4 - 24 = -12$$

(*f • g*)(*x*)

To multiply the two functions, distribute the terms of one function over the terms of the second function:

$$(f \cdot g)(x) = f(x) \cdot g(x)$$
$$= (2x^2 + 5x - 12)(3x + 12)$$
$$= 2x^2(3x) + 2x^2(12) + 5x(3x) + 5x(12) - 12(3x) - 12(12)$$
$$= 6x^3 + 24x^2 + 15x^2 + 60x - 36x - 144$$
$$= 6x^3 + 39x^2 + 24x - 144$$

Double-check:

$f(2) = 6$ and $g(2) = 18$

$6 \times 18 = 108$

$(f \cdot g)(x) = 6(2)^3 + 39(2)^2 + 24(2) - 144 = 48 + 156 + 48 - 144 = 108$

$$\left(\frac{f}{g}\right)(x)$$

$$\left(\frac{f}{g}\right)(x) = \frac{f(x)}{g(x)}$$
$$= \frac{2x^2 + 5x - 12}{3x + 12}$$
$$= \frac{(2x - 3)\cancel{(x + 4)}}{3\cancel{(x + 4)}}$$
$$= \frac{2x - 3}{3}, x \neq -4$$

If $x = -4$, the denominator is $3(-4) + 12 = 0$ and $\frac{f(x)}{g(x)}$ undefined. Therefore, $x \neq -4$.

Double-check:

$f(2) = 6$ and $g(2) = 18$

$$\frac{6}{18} = \frac{1}{3}$$

$$\left(\frac{f}{g}\right)(x) = \frac{2(2) - 3}{3} = \frac{1}{3}$$

Another way of combining two functions is to composite them, as in the following example.

Composite Functions: Let $f(x) = 3x - 2$ and $g(x) = x + 6$. Find $f(g(2))$.

Step 1: Evaluate $g(x)$ first.

Step 2: Use the output of g as the input of f and evaluate $f(x)$.

Solution and Explanation: A function in the form $f(g(x))$ or $(f \circ g)(x)$, which is read "f of g of x", is called a composite function.

To find $f(g(2))$, we first evaluate $g(2)$ and get 8 as the output:
$g(2) = 2 + 6 = 8$.

The output of $g(2)$ then becomes the input for $f(x)$:
$f(8) = 3(8) - 2 = 22$.

Therefore, $f(g(2)) = 22$.

If $f(g(x)) = x$ and $g(f(x)) = x$, then $f(x)$ and $g(x)$ are inverses of each other. This means that $f(x)$ consists of the ordered pairs (x,y) and $g(x)$ consists of ordered pairs (y,x). We'll explore inverse functions in more detail in the next problem.

Inverse Functions: Let $f(x) = 2x + 1$. Find $f^{-1}(x)$.

Step 1: Replace $f(x)$ with y in the original equations.

Step 2: Interchange x and y.

Step 3: Solve for y.

Step 4: Replace y with $f^{-1}(x)$.

Solution and Explanation:

$$f(x) = 2x + 1$$
$$y = 2x + 1$$
$$x = 2y + 1$$
$$x - 1 = 2y$$
$$\frac{x-1}{2} = y$$
$$\frac{x-1}{2} = f^{-1}(x)$$

To confirm that f^{-1} is the inverse of f, we evaluate the composition of f and f^{-1}.

$$f\left(f^{-1}(x)\right) = f\left(\frac{x-1}{2}\right) = 2\left(\frac{x-1}{2}\right) + 1 = x - 1 + 1 = x$$

$$f^{-1}\left(f(x)\right) = f^{-1}(2x+1) = \frac{(2x+1)-1}{2} = \frac{2x}{2} = x$$

Since both compositions equal x, f and f^{-1} are inverses of one another. A point (a,b) on the graph of f will be a point (b,a) on the graph of f^{-1}.

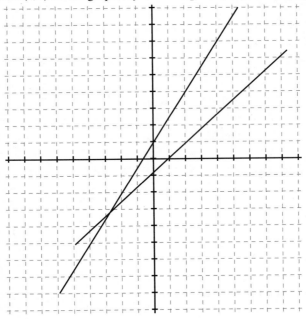

CHAPTER QUIZ

1. Graph $f(x) = -2x^2 + 3$.

(A)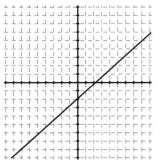

(B)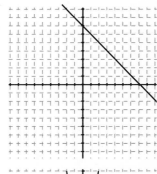

(C)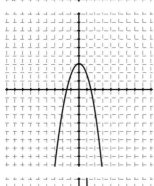

(D)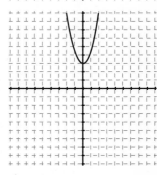

(E)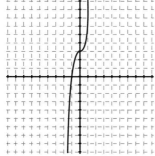

2. Which graph below would most likely be the graph of $f(x) = x^3 + 4x^2 - 1$?

(A)

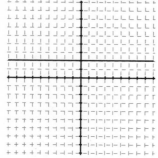

(B)

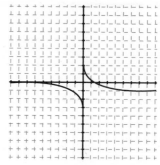

(C)

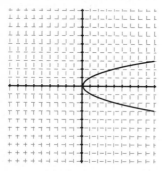

(D)

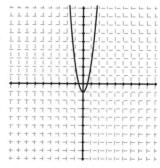

(E)

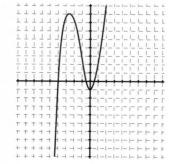

3. Find the zeros of
   $f(x) = x^3 - 2x^2 - 8x$.

   (A) $x = -4, x = -2$

   (B) $x = 4, x = 2$

   (C) $x = 0, x = -4, x = 2$

   (D) $x = 0, x = 4, x = -2$

   (E) $x = 0, x = 4, x = 2$

4. Find the zeros of
   $f(x) = x^4 - 13x^2 + 36$.

   (A) $x = 2, x = -2, x = 3, x = -3$

   (B) $x = -2, x = -3$

   (C) $x = 2, x = 3$

   (D) $x = 0, x = -2, x = 3$

   (E) $x = 0, x = 2, x = -3$

5. Let $f(x) = x^2 - 1$ and
   $g(x) = x^2 - x - 2$. For what value
   of x is $\left(\dfrac{f}{g}\right)(x)$ undefined?

   (A) $x = -2$

   (B) $x = -1$

   (C) $x = 0$

   (D) $x = 1$

   (E) $x = 2$

6. Let $f(x) = x - 3$ and $g(x) = x^2 + 1$.
   For what value of x is $\left(\dfrac{f}{g}\right)(x)$
   undefined?

   (A) 0

   (B) -1

   (C) 1

   (D) $\emptyset$

   (E) None of the above.

7. Let $f(x) = 3x - 1$ and $g(x) = x + 5$.
   Find $f(g(2))$.

   (A) 18

   (B) 19

   (C) 20

   (D) 21

   (E) 22

8. Let $f(x) = x^2$ and $g(x) = x - 2$.
   Find $(g \circ f)(2)$.

   (A) 0

   (B) 1

   (C) 2

   (D) 3

   (E) 4

9. Find the inverse of $f(x) = x^3 - 5$.

   (A) $y^3 - 5$

   (B) $x + 5$

   (C) $\sqrt[3]{y + 5}$

   (D) $\sqrt[3]{x + 5}$

   (E) $\sqrt[3]{x - 5}$

10. Which of the following could be the graphs of $f(x) = x^3 - 5$ and its inverse?

(A)

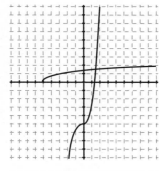

(B)

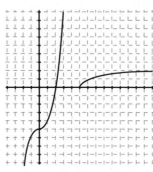

(C)

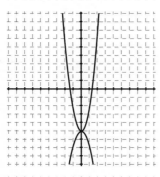

(D)

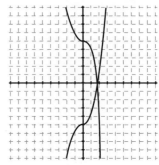

(E)

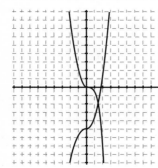

Answers and Explanations

1. C

Recall that the graph of a quadratic function (a function of degree 2) is a parabola, so neither (A), (B), or (E) can be the graph of $f(x) = -2x^2 + 3$. To graph the parabola, make a table of values for x and $f(x)$.

x	$f(x)$	$(x, f(x))$
-2	$-2(-2)^2 + 3 = -5$	$(-2,-5)$
-1	$-2(-1)^2 + 3 = 1$	$(-1,1)$
0	$-2(0)^2 + 3 = 3$	$(0, 3)$
1	$-2(1)^2 + 3 = 1$	$(1,1)$
2	$-2(2)^2 + 3 = -5$	$(2,-5)$

Plot the points and you get answer (C). Note the differences between the graph of $f(x) = x^2$ and the graph of $f(x) = -2x^2 + 3$. The "+ 3" indicates that the parabola is shifted up 3 units and the negative leading coefficient (-2) indicates that the parabola faces downwards.

The function for the graph in (D) is $f(x) = x^2 + 3$. It is also shifted up 3 units but faces upwards because its term of highest degree (x^2) is positive.

2. E

Choice (A) cannot be the graph because it is a constant function. This means that the value of $f(x)$ is always the same number regardless of the value of x. In this case, $f(x) = 2$. Choice (B) resembles the graph of $f(x) = \frac{1}{x}$ and (D) is the graph of a quadratic function. Recall that a function is a relationship between x and y such that each value of x corresponds to one and only one value of y. Therefore, (C) is not the graph of a function.

3. D

Write the equation $x^3 - 2x^2 - 8x = 0$ and solve by factoring.

$$x^3 - 2x^2 - 8x = 0$$
$$x\left(x^2 - 2x - 8\right) = 0$$
$$x(x - 4)(x + 2) = 0$$

Recall the Principle of Zero Products, and set the monomial and binomials equal to zero.

$$x = 0 \quad \text{or} \quad \begin{matrix} x - 4 = 0 \\ x = 4 \end{matrix} \quad \text{or} \quad \begin{matrix} x + 2 = 0 \\ x = -2 \end{matrix}$$

Double-check:

$$f(0) = (0)^3 - 2(0)^2 - 8(0) = 0$$
$$f(4) = (4)^3 - 2(4)^2 - 8(4) = 64 - 32 - 32 = 0$$
$$f(-2) = (-2)^3 - 2(-2)^2 - 8(-2) = -8 - 8 + 16 = 0$$

4. A

Set the function equal to zero and solve by factoring.

$$x^4 - 13x^2 + 36 = 0$$
$$\left(x^2 - 4\right)\left(x^2 - 9\right) = 0$$
$$(x + 2)(x - 2)(x + 3)(x - 3) = 0$$

$$\begin{matrix} x + 2 = 0 \\ x = -2 \end{matrix} \quad \text{or} \quad \begin{matrix} x - 2 = 0 \\ x = 2 \end{matrix} \quad \text{or} \quad \begin{matrix} x + 3 = 0 \\ x = -3 \end{matrix} \quad \text{or} \quad \begin{matrix} x - 3 = 0 \\ x = 3 \end{matrix}$$

Double-check:

$$f(-2) = (-2)^4 - 13(-2)^2 + 36 = 16 - 52 + 36 = 26$$
$$f(2) = (2)^4 - 13(2)^2 + 36 = 16 - 52 + 36 = 0$$
$$f(-3) = (-3)^4 - 13(-3)^2 + 36 = 81 - 117 + 36 = 0$$
$$f(3) = (3)^4 - 13(3)^2 + 36 = 81 - 117 + 36 = 0$$

5. E

$$\frac{f(x)}{g(x)} = \frac{x^2 - 1}{x^2 - x - 2}$$
$$= \frac{(x + 1)(x - 1)}{(x - 2)(x + 1)}$$
$$= \frac{x - 1}{x - 2}, x \neq 2$$

The function is undefined when $x - 2 = 0$. Therefore, it is undefined when $x = 2$.

6. D

$$\frac{f(x)}{g(x)} = \frac{x-3}{x^2+1}$$

We cannot factor the denominator because there are no two integers the product and sum of which both equal positive one. If we set the denominator equal to zero, we get

$$x^2 + 1 = 0$$

$$x^2 = -1$$

This is not possible since x^2 must always be positive. Therefore, there are no values of x for which the function will be undefined.

7. C

To solve composite functions, we evaluate the inner function first. The output of the inner function becomes the input for the outer function. In this case, $g(x)$ is the inner function and $f(x)$ is the outer function.

First, evaluate the function for $g(2)$.

$$g(2) = 2 + 5 = 7.$$

The output 7 becomes the input for $f(x)$.

Evaluate the function for $f(7)$.

$$f(7) = 3(7) - 1 = 20$$

8. C

$(g \circ f)(x) = g(f(x))$. Be sure that you evaluate $f(x)$ first, and then $g(x)$.

$$f(2) = (2)^2 = 4$$

The output of $f(2)$ becomes the input for $g(x)$.

$$g(4) = 4 - 2 = 2$$

9. D

Replace $f(x)$ with y:

$$y = x^3 - 5$$

Interchange the variables:

$$x = y^3 - 5$$

Solve for y:

$$x + 5 = y^3$$

$$\sqrt[3]{x+5} = y$$

Replace y with $f^{-1}(x)$:

$$\sqrt[3]{x+5} = f^{-1}(x)$$

Check that $f(f^{-1}(x)) = f^{-1}(f(x)) = x$:

$$f\left(f^{-1}(x)\right) = f\left(\sqrt[3]{x+5}\right) = \left(\sqrt[3]{x+5}\right)^3 - 5 = (x+5) - 5 = x$$

$$f^{-1}\left(f(x)\right) = f^{-1}\left(x^3 - 5\right) = \sqrt[3]{\left(x^3 - 5\right) + 5} = \sqrt[3]{x^3} = x$$

10. A

Graph $f(x) = x^3 - 5$ by creating a table of values for x and $f(x)$.

x	$f(x)$	$(x, f(x))$
-2	$(-2)^3 - 5 = -13$	$(-2, -13)$
-1	$(-1)^3 - 5 = -6$	$(-1, -6)$
0	$(0)^3 - 5 = -5$	$(0, -5)$
1	$(1)^3 - 5 = -4$	$(1, -4)$
2	$(2)^3 - 5 = 3$	$(2, 3)$

The coordinates of the f^{-1} would be the reverse of the coordinates of $f(x)$. For example, the ordered pair $(0,5)$ for $f(x)$ would be $(5,0)$ for f^{-1}.

Plotting the coordinates of f and f^{-1} gives us answer (A).

CHAPTER 4

Linear Equations

WHAT IS A LINEAR EQUATION?

A linear equation is any equation with a graph that is a straight line. Recall from Chapter 3 that the graph of $y = x$ is a straight line that passes through the origin of the xy plane. You will see in this chapter that the equation of a line can be written in several different ways. One thing these equations have in common is that the degree of x is always 1.

$y = \dfrac{1}{x}$ is not the equation of a line because the degree of x is -1: $\dfrac{1}{x} = x^{-1}$.

$y = \sqrt{x}$ is also not an equation of a line because $\sqrt{x} = x^{\frac{1}{2}}$.

CONCEPTS TO HELP YOU

1. Slope of a line: The steepness of a line is called its slope. The slope of a line is usually denoted by the letter m and its formula is $m = \dfrac{y_2 - y_1}{x_2 - x_1}$ where (x_1, y_1) and (x_2, y_2) are two points on the line.

    The greater the steepness of a line, the greater is its slope.

    Lines that go upward from left to right have a positive slope and lines that go downward from left to right have a negative slope.

    Parallel lines have the same slope.

    The slopes of perpendicular lines are negative reciprocals.

    A horizontal line has slope of 0.

    A vertical line does not have a slope.

2. Standard form of a linear equation: $Ax + By = C$, where A, B, and C are integer coefficients and A is positive.

3. Slope-intercept equation of a line: $y = mx + b$, where m is the slope of the line and b is the y-coordinate of the line's y-intercept. The y-intercept is the point at which the line crosses the y-axis.

4. Point-slope equation of a line: $y - y_0 = m(x - x_0)$, where m is the slope of the line and (x_0, y_0) is a point on the line.

STEPS YOU NEED TO REMEMBER

1. Identify the given values.

If you are given that the slope of a line is 3, then $m = 3$.

If you are given that the y-intercept of a line is 3, then $b = 3$.

If you are given two points on a line, then those two points are (x_1, y_1) and (x_2, y_2).

If you know only one point on the line, then that point is (x_0, y_0).

2. Determine the best form of the equation to use.

Use the slope-intercept equation $y = mx + b$ if you are given the slope and the y-intercept.

Use the point-slope equation $y - y_0 = m(x - x_0)$ if you are given the slope and a point on the line that is not the y-intercept.

3. Solve for m.

If you have not been given the slope but you know two points (x_1, y_1) and (x_2, y_2) on the line, you need to solve for m.

4. Plug the values into the equation.

If you know the values of m and b, plug them into the equation $y = mx + b$.

If you know the values of m and a point (x_0, y_0) on the line, plug them into the equation $y - y_0 = m(x - x_0)$.

COMMON LINEAR EQUATION QUESTIONS

Slope of a Line: Find the slope of the line that contains the points:

(−1, −1) and (1,3)

(−1,4) and the origin

(−2, 1) and (4,1)

(−2,1) and (−2,−1)

Step 1: Plug the coordinates into the formula $m = \dfrac{y_2 - y_1}{x_2 - x_1}$.

Step 2: Solve the formula for those coordinates.

Solution and Explanation:

Let (−1, −1) be the ordered pair (x_1, y_1) and let (1,3) be the ordered pair (x_2, y_2).

$$m = \frac{y_2 - y_1}{x_2 - x_1} = \frac{3 - (-1)}{1 - (-1)} = \frac{4}{2} = 2$$

It does not matter which ordered pair you designate as (x_1, y_1) and which one you designate as (x_2, y_2). The slope will be the same.

If (1,3) is (x_1, y_1) and (−1, −1) is (x_2, y_2), then

$$m = \frac{y_2 - y_1}{x_2 - x_1} = \frac{-1 - 3}{-1 - 1} = \frac{-4}{-2} = 2 \ .$$

Let's graph the line to confirm that the slope is 2 anywhere on the line.

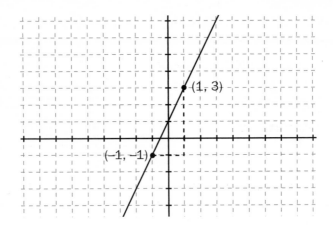

As you can see, when we move from $(-1,-1)$ to $(1,3)$, we move 2 spaces to the right and up 4 spaces. In other words, the change in y is 4 and the change in x is 2.

Now let's pick two other points on the line.

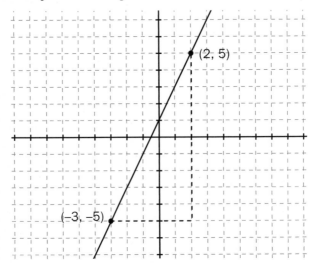

To get from $(-3,-5)$ to $(2,5)$, we move 5 spaces to the right and 10 spaces up, so the change in x is 5 and the change in y is 10:

$$m = \frac{y_2 - y_1}{x_2 - x_1} = \frac{10}{5} = 2$$

The slope of the line is always 2, regardless of where you are on the line.

Let $(-1,4)$ be (x_1,y_1) and let the origin, whose coordinates are $(0,0)$, be (x_2,y_2).

$$m = \frac{y_2 - y_1}{x_2 - x_1} = \frac{0-4}{0-(-1)} = -4$$

The slope of this line is negative. When we graph it, we see that the line points downwards to the right.

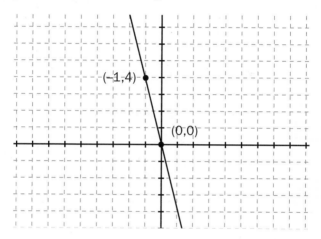

To get from $(-1,4)$ to $(0,0)$, we move 1 space to the right and 4 spaces down.

Remember that lines with negative slopes point downwards to the right and lines with positive slopes point upwards to the right.

Let $(-2, 1)$ be (x_1,y_1) and let $(4,1)$ be (x_2,y_2).

$$m = \frac{y_2 - y_1}{x_2 - x_1} = \frac{1-1}{4-(-2)} = \frac{0}{6} = 0$$

When we plot the points, we see that the graph is a horizontal line. In fact, the slope of all horizontal lines is zero.

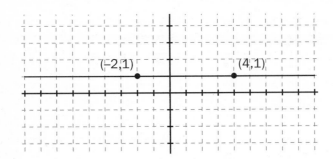

From the graph, we can see that we move horizontally to get from one point to the other but we do not move vertically at all. There is no change in y.

Let $(-2,1)$ be (x_1,y_1) and let $(-2,-1)$ be (x_2,y_2).

$$m = \frac{y_2 - y_1}{x_2 - x_1} = \frac{-3 - 3}{3 - 3} = \frac{-6}{0} = \text{undefined}$$

The slope is undefined because a fraction is always undefined when the denominator is zero.

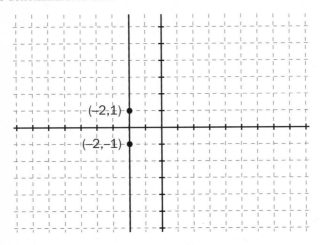

The graph of this line is a vertical line. The slope of all vertical lines is undefined or nonexistent.

You have been practicing how to calculate the slope of a line given two points on the line. Now you will write linear equations of various forms.

Equation of a Line Given the Slope and *y*-Intercept: Write the equation of a line with slope –2 and *y*-intercept 3 and graph the line.

Step 1: Choose the slope-intercept form $y = mx + b$, where m is the slope of the line and b is the y-coordinate of the line's y-intercept. The y-intercept is the point at which the line crosses the y-axis.

Step 2: Plug in the values of the slope and y-intercept into the equation.

Step 3: Solve the equation for the x-intercept.

Step 4: Plot the x and y intercepts on a coordinate plane and draw a line through the points.

Solution and Explanation: Because you are given the slope and the y-intercept, you can write the slope-intercept equation of the line. The slope-intercept form of a line is $y = mx + b$ where m is the slope of the line and b is the y-intercept. Therefore, the equation of the line is $y = -2x + 3$.

Now that we have an equation, we can graph the line. We already know one point on the line—the coordinates of the y-intercept, which is (0,3). We can verify this by plugging in 0 for the value of x into the equation:

$$y = -2x + 3$$
$$= -2(0) + 3$$
$$= 3$$

All we need now is another point on the line. We can choose any value for x, substitute it into the equation and solve for y. It is a good idea, however, to plot the x-intercept as the second point on the line.

The x-intercept is the point at which the line crosses the x-axis. At this point, $y = 0$.

$$0 = -2x + 3$$
$$-3 = -2x$$
$$\frac{3}{2} = x$$

The coordinate of the x-intercept is $\left(\frac{3}{2}, 0\right)$. Plot this point and the y-intercept and you get the graph below:

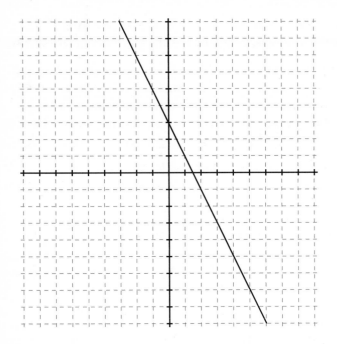

Next, you will write an equation in point-slope form and then convert the point-slope form to standard form.

Equation of a Line Given the Slope and a Point on the Line: If a line contains the points (–2,5) and has slope –1, write the equation of the line in standard form.

Step 1: Choose the point-slope form $y - y_0 = m(x - x_0)$, where m is the slope of the line and (x_0, y_0) is a point on the line.

Step 2: Plug the appropriate values into the equation.

Step 3: Simplify the equation.

Step 4: Rearrange the equation into the standard form $Ax + By = C$, where A, B, and C are integer coefficients and A is positive.

> **Solution and Explanation:** Because we are given a point on the line and the slope of the line, we first write the equation in point-slope form with $m = -1$ and $(-2,5)$ as the values of (x_0, y_0).
>
> $$y - y_0 = m\left(x - x_0\right)$$
> $$y - 5 = -1\left(x - (-2)\right)$$
> $$y - 5 = -x - 2$$
> $$y = -x + 3$$
>
> Verify that this equation is correct by substituting $(-2,5)$ into $y = -x + 3$.
>
> $$y = -x + 3$$
> $$5 = -(-2) + 3$$
> $$5 = 2 + 3$$
>
> Now we need to rearrange $y = -x + 3$ into the standard form $Ax + By = C$.
>
> $$y = -x + 3$$
> $$y + x = -x + x + 3$$
> $$x + y = 3$$
>
> In standard form, $A = 1$, $B = 1$, and $C = 3$.

So far, we have dealt with only one line at a time. The following problems concern two lines that are either parallel or perpendicular to each other.

Slope of Parallel Lines: The lines below are parallel. What is the slope of *f* and *g*?

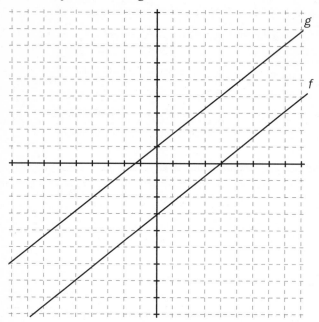

Step 1: Calculate the slope of *f* using the *x* and *y* intercepts.

Step 2: Set the slope of *g* equal to the slope of *f*.

> **Solution and Explanation:** Two lines are said to be parallel if they lie in the same plane and, if extended infinitely in either direction, they never intersect. Parallel lines have the same slope.
>
> We can tell by looking at the graph that line *f* has a positive slope and that it crosses the *x*-axis at 4 and the *y*-axis at –3, so two points on line *f* are (4,0) and (0,–3).
>
> $$m = \frac{y_2 - y_1}{x_2 - x_1} = \frac{-3 - 0}{0 - 4} = \frac{-3}{-4} = \frac{3}{4}$$
>
> The slope of *f* is $\frac{3}{4}$.
>
> Since line *g* is parallel to line *f*, the slope of *g* = the slope of *f* = $\frac{3}{4}$.

Slope of Perpendicular Lines: The lines below are perpendicular. What is the slope of *g*?

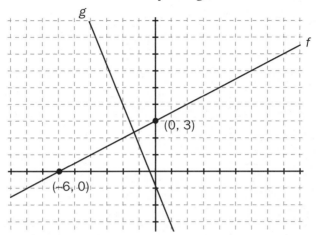

Step 1: Calculate the slope of *f*.

Step 2: For all real numbers *a* and *b*, the negative reciprocal of $\frac{a}{b}$ is $-\frac{b}{a}$. Write the reciprocal of the slope of *f* by flipping its values around. If the slope of *f* is positive, make the reciprocal negative. If the slope of *f* is negative, make the reciprocal positive.

Solution and Explanation: Two lines are said to be perpendicular if they meet at a right angle. The slopes of perpendicular lines are negative reciprocals.

Because we know two points on the line of *f*, we can calculate the slope of *f*.

$$m = \frac{y_2 - y_1}{x_2 - x_1} = \frac{3 - 0}{0 - (-6)} = \frac{3}{6} = \frac{1}{2}$$

The negative reciprocal of $\frac{a}{b}$ is $-\frac{b}{a}$, so the negative reciprocal of $\frac{1}{2}$ is $-\frac{2}{1}$ or -2.

Therefore, the slope of *g* is -2.

Looking at the graph, we note that *f* slants upwards toward the right so its slope should be positive. Line *g* slants downwards to the right so its slope should be negative.

CHAPTER QUIZ

1. Which of the graphs below is the graph of a line with a slope of –3 and that passes through the point (2,–3)?

(A) 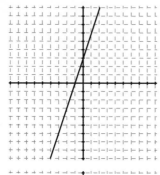   (B)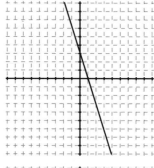

(C) 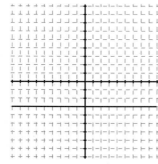   (D)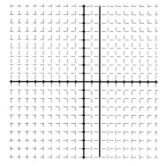

(E)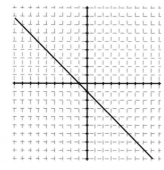

2. Two lines are shown below, one with a slope of 7 and the other with a slope of $\frac{1}{4}$. Which of the graphs below could be the graphs of the two lines?

(A)
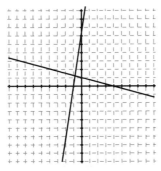

(B)
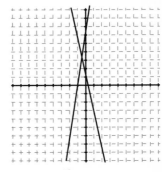

(C)
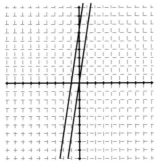

(D)
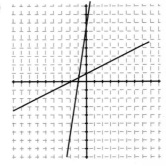

(E)
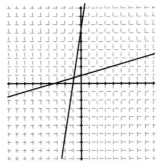

3. What is the equation in slope-intercept form for a line that has a slope of $\frac{3}{4}$ and passes through $(0,-7)$?

(A) $y = \frac{3}{4}x + 7$

(B) $y - 7 = \frac{3}{4}x$

(C) $y + 7 = \frac{3}{4}x$

(D) $y = \frac{3}{4}x - 7$

(E) $\frac{3}{4}x - y = 7$

4. Which is the equation of a line that contains $(2,0)$ and $(-5,8)$?

(A) $y - 8 = \frac{8}{-7}(x + 5)$

(B) $y - 0 = \frac{8}{-7}(x - 2)$

(C) $y = \frac{8}{7}x + \frac{16}{7}$

(D) Both A and B

(E) Both B and C

5. Which of the following is an equation for the line that passes through $(-2,1)$ and is perpendicular to the line whose equation is $y = 5x + 3$.

(A) $y - 1 = -\frac{1}{5}x - \frac{2}{5}$

(B) $y = -\frac{1}{5}x + \frac{3}{5}$

(C) $\frac{1}{5}x + y = \frac{3}{5}$

(D) $x + 5y = 3$

(E) All of the above.

6. Write $\frac{2}{3}x = -6y + 1$ in standard form.

(A) $\frac{2}{3}x + 6y = 1$

(B) $-\frac{2}{3}x - 6y = -1$

(C) $6y + \frac{2}{3}x = 1$

(D) $6y - \frac{2}{3}x = 1$

(E) None of the above.

7. Which of the following could be the equation of a line parallel to the line $y = 3x - 2$?

(A) $y = 3x - 1$

(B) $y = 3x + 2$

(C) $y = 3x - 3$

(D) $y = 3x + 4$

(E) All of the above.

8. Which of the following is the equation of a line that contains the point $(-2,5)$ and is parallel to the line $3x - 2y = 6$?

(A) $y = \frac{3}{2}x + 8$

(B) $y = -\frac{2}{3}x + 8$

(C) $y = \frac{2}{3}x + 8$

(D) $y = \frac{2}{3}x - 8$

(E) $y = -\frac{2}{3}x - 8$

9. What is the slope of the line
   $y = -3$?
   - (A) -3
   - (B) $-\dfrac{1}{3}$
   - (C) 0
   - (D) $\dfrac{1}{3}$
   - (E) 3

10. What is the slope of the line
    $x = 1$?
    - (A) 1
    - (B) -1
    - (C) 0
    - (D) undefined
    - (E) there is not enough information

Answers and Explanations

1. B

The slope of the line is negative, so we know that the line must point downwards to the right. This automatically eliminates choice (A) because its line points upwards, indicating that it has a positive slope. Also, we can see that it does not pass through the point (2,–3). The line in choice (D) passes through (2,–3) but it is a vertical line, so it has no slope. The line in choice (E) is a horizontal line, so its slope is 0.

Both the lines in (B) and (C) have a negative slope and appear to pass through the point (2,–3). To determine which one has a slope of –3, we need to choose two points on each line.

Line (B) crosses the y-axis at 3 and the x-axis at 1, so it contains the points (0,3) and (1,0). Therefore, its slope is $m = \dfrac{y_2 - y_1}{x_2 - x_1} = \dfrac{0 - 3}{1 - 0} = -3$.

Another way of calculating the slope is to draw a right triangle with (0,3) and (1,0) as vertices of the triangle and the distance from (0,3) and (1,0) as the hypotenuse.

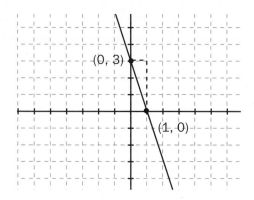

The base of the right triangle denotes the change in x and the height denotes the change in y. To get from $(0,3)$ to $(1,0)$, we move one unit to the right and 3 units down. Moving one unit horizontally means that the change in x is 1. Moving horizontally to the right means the change is a positive 1. Moving vertically 3 units down means that the change in y is -3. Therefore, the slope is $\frac{-3}{1} = -3$.

Let's see what the slope of line (D) is. The line crosses the x-axis at -1 and the y-axis at -1 as well, so the line contains the points $(-1,0)$ and $(0,-1)$. Therefore, its slope is $m = \frac{y_2 - y_1}{x_2 - x_1} = \frac{-1-0}{0-(-1)} = -1$.

2. E

In choice (A), one of the lines passes through $(-1,0)$ and $(0,7)$ so its slope is $m = \frac{7-0}{0-(-1)} = 7$. In fact, all the answer choices have that line in common, so we know that they all contain a line with a slope of 7. The other line must have a negative slope because it points downwards. Because it passes through $(0,1)$ and $(4,0)$, its slope is $m = \frac{0-1}{4-0} = \frac{-1}{4}$.

We know that choice (B) has a line with a slope of 7. We also know, simply by looking at the graph, that the other line cannot have a slope of $\frac{1}{4}$ because it points downwards.

The lines in choice (C) are parallel, so they both have the same slope of 7.

Notice that, in choices (D) and (E), both lines have positive slopes but one is relatively steeper than the other. The steeper the line, the greater is its slope. The line with a slope of 7 is closer to the vertical position and the other line is closer to the horizontal position. We need to determine which graph contains the line with a slope of $\frac{1}{4}$.

We can calculate the slope in choice (D) visually or using the slope formula. Visually, we see that to get from the x-intercept to the y-intercept, we need to move 2 units to the right and 1 unit up, so the slope of this line is $\frac{1}{2}$. Using the slope formula and the coordinates (−2,0) and (0,1), we get $m = \frac{1-0}{0-(-2)} = \frac{1}{2}$.

The answer, therefore, must be (E). The line with the smaller slope passes through the points (−4,0) and (0,1) so its slope is $m = \frac{1-0}{0-(-4)} = \frac{1}{4}$.

3. D

The slope-intercept form of a line is $y = mx + b$, where m is the slope of the line and b is the y-coordinate of the line's y-intercept.

We know that $m = \frac{3}{4}$. The point (0,−7) is the y-intercept of the line, so b is −7. Plug these values into the $y = mx + b$ and you get $y = \frac{3}{4}x - 7$.

Choices (C) and (E) are equations of the same line in point-slope form and standard form.

The point-slope form of a line is $y - y_0 = m(x - x_0)$, where m is the slope of the line and (x_0, y_0) is a point on the line. Plug the values into the equation and you get

$$y - y_0 = m\left(x - x_0\right)$$
$$y - (-7) = \frac{3}{4}(x - 0)$$
$$y + 7 = \frac{3}{4}x$$

The standard form of a line is $Ax + By = C$, where A, B, and C are integer coefficients and A is positive. We can rearrange the point-slope equation to arrive at the standard form.

$$y + 7 = \frac{3}{4}x$$
$$y - y + 7 = \frac{3}{4}x - y$$
$$7 = \frac{3}{4}x - y$$

4. D

We know two points on the line so we can calculate the slope and plug the values into the point-slope form to write the equation of the line.

$$m = \frac{8-0}{-5-2} = \frac{8}{-7}$$

It does not matter whether you choose (2,0) or (−5,8) to plug into the point-slope equation because they are on the same line.

$$y - y_0 = m(x - x_0)$$

$$y - y_0 = m(x - x_0) \qquad\qquad y - 8 = \frac{8}{-7}(x - (-5))$$

$$y - 0 = \frac{8}{-7}(x - 2) \quad \text{or} \quad y - 8 = \frac{8}{-7}x - \frac{40}{7}$$

$$y = \frac{8}{-7}x + \frac{16}{7} \qquad\qquad y - 8 + 8 = \frac{8}{-7}x - \frac{40}{7} + 8$$

$$y = \frac{8}{-7}x + \frac{16}{7}$$

Choice (A) is the point-slope form of the equation using (−5,8) as the point and choice (B) is the point-slope form of the equation using (2,0) as the point. Therefore, both (A) and (B) are correct.

Choice (C) resembles the slope-intercept form of the line except that the slope in (C) is positive.

5. E

The given equation $y = 5x + 3$ is the slope-intercept form of the line where $m = 5$ and $b = 3$. The slope of a line perpendicular to the given line will be the negative reciprocal of 5. Recall that the negative reciprocal of $\frac{a}{b}$ is $-\frac{b}{a}$, so the slope of the perpendicular line is $-\frac{1}{5}$.

When we plug $-\frac{1}{5}$ and the point $(-2,1)$ into the point-slope equation of a line and simplify, we get choice (A).

$$y - y_0 = m\left(x - x_0\right)$$
$$y - 1 = -\frac{1}{5}\left(x - (-2)\right)$$
$$y - 1 = -\frac{1}{5}x - \frac{2}{5}$$

When we simplify choice (A) even further, we get choice (B).

$$y - 1 = -\frac{1}{5}x - \frac{2}{5}$$
$$y - 1 + 1 = -\frac{1}{5}x - \frac{2}{5} + 1$$
$$y = -\frac{1}{5}x - \frac{2}{5} + 1$$
$$y = -\frac{1}{5}x + \frac{3}{5}$$

When we rewrite the equation in choice (B) in standard form, we get choice (C).

$$y = -\frac{1}{5}x + \frac{3}{5}$$
$$y + \frac{1}{5}x = -\frac{1}{5}x + \frac{1}{5}x + \frac{3}{5}$$
$$\frac{1}{5}x + y = \frac{3}{5}$$

When we simplify the standard form by multiplying through by 5, we get choice (D).

$$\frac{1}{5}x + y = \frac{3}{5}$$
$$\cancel{5} \cdot \frac{1}{\cancel{5}}x + 5 \cdot y = \cancel{5} \cdot \frac{3}{\cancel{5}}$$
$$x + 5y = 3$$

Therefore, choices (A), (B), (C), and (D) are all equations of the line.

6. A

Rearrange $\frac{2}{3}x = -6y+1$ into the form $Ax + By = C$, where A, B, and C are integer coefficients and A is positive.

$$\frac{2}{3}x = -6y+1$$

$$\frac{2}{3}x+6y = -6y+6y+1$$

$$\frac{2}{3}x+6y = 1$$

If we had subtracted $\frac{2}{3}x$ from both sides instead of added –6y to both sides, we would have gotten choice (B).

$$\frac{2}{3}x = -6y+1$$

$$\frac{2}{3}x-\frac{2}{3}x = -6y-\frac{2}{3}x+1$$

$$0-1 = -6y-\frac{2}{3}x+1-1$$

$$-1 = -6y-\frac{2}{3}x$$

$$-\frac{2}{3}x-6y = -1$$

Part of the definition of the standard form is that A is positive, so choice (B) is not the standard form of the equation. Neither are choices (C) and (D) because they are both in the form $By + Ax = C$.

7. E

The slopes of parallel lines are equal. The slope of the line $y = 3x - 2$, where $m = 3$ and $b = -2$, is 3. Therefore, the slope of a parallel line is also 3.

All the equations in the answer choices are in the form $y = mx + b$ and their values of m all equal 3. Therefore, all of them are equations of lines parallel to $y = 3x - 2$. The only difference between the lines is where they intercept the x- and y-axis.

8. A

Rearranging the equation $3x - 2y = 6$ into slope-intercept form will give us the slope of the line.

$$3x - 2y = 6$$
$$3x - 3x - 2y = 6 - 3x$$
$$-2y = 6 - 3x$$
$$\frac{-2y}{-2} = \frac{6}{-2} - \frac{3x}{-2}$$
$$y = -3 + \frac{3x}{2}$$
$$y = \frac{3}{2}x - 3$$

A line parallel to this will also have slope $\frac{3}{2}$. This line passes through the point $(-2,5)$ so we begin by writing the point-slope equation of the line.

$$y - y_0 = m(x - x_0)$$
$$y - 5 = \frac{3}{2}(x - (-2))$$
$$y - 5 = \frac{3}{2}x + 3$$
$$y = \frac{3}{2}x + 8$$

We could have answered the question correctly even without writing out the equation because (A) is the only answer choice that gives $\frac{3}{2}$ as the slope.

9. C

The graph of $y = -3$ is a horizontal line that passes through the y-axis at -3. Keep in mind that the graph of an equation in the form $y = k$ where k is a constant is always a horizontal line. It does not matter what the value of x is because the value of y is fixed at k.

The slope of a horizontal line is always 0, so the slope of $y = -3$ is zero.

10. D

The graph of $x = 1$ is a vertical line that passes through the x-axis at 1 because all ordered pairs on the line are $(y,1)$. A line in the form $x = k$ where k is a constant is always a vertical line.

A vertical line has no slope, or its slope is undefined.

CHAPTER 5

Trigonometric Functions

WHAT ARE TRIGONOMETRIC FUNCTIONS?

Trigonometric functions are functions of angles. They involve sine, cosine, tangent, and their reciprocals. On a circle centered at the origin of an xy plane with a radius of 1 and an angle θ,

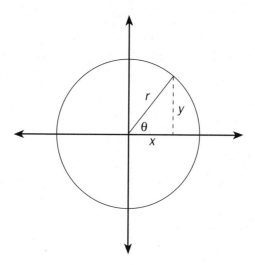

the trigonometric functions are:

$\sin\theta = \dfrac{y}{r}$	$\csc\theta = \dfrac{1}{\sin\theta} = \dfrac{r}{y}$
$\cos\theta = \dfrac{x}{r}$	$\sec\theta = \dfrac{1}{\cos\theta} = \dfrac{r}{x}$
$\tan\theta = \dfrac{\sin\theta}{\cos\theta} = \dfrac{y}{x}$	$\cot\theta = \dfrac{1}{\tan\theta} = \dfrac{x}{y}$

CONCEPTS TO HELP YOU

1. Graphs of trigonometric functions: The graphs of the trigonometric functions are called periodic functions because they repeat themselves after a fixed period of time.

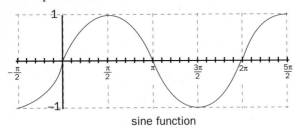

sine function

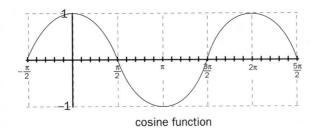

cosine function

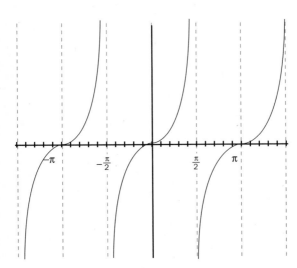

tangent function

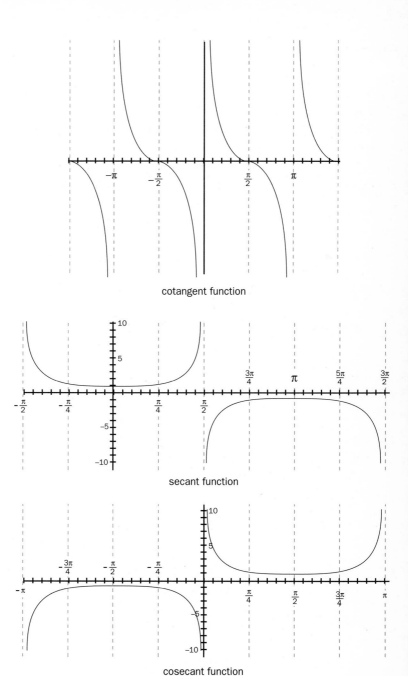

cotangent function

secant function

cosecant function

2. Degrees and radians: In trigonometry, angles are measured in terms of degrees or radians. Radians are expressed in terms of π. There are 360° per 2π radians. A straight angle, which is 180°, is equal to π radians.

3. Trigonometric values for common angles: To solve trigonometric functions, you will need to know the following trigonometric values for common angles.

Angle (degrees)	Angle (radians)	Sine	Cosine
0°	0	0	1
30°	$\dfrac{\pi}{6}$	$\dfrac{1}{2}$	$\dfrac{\sqrt{3}}{2}$
45°	$\dfrac{\pi}{4}$	$\dfrac{\sqrt{2}}{2}$	$\dfrac{\sqrt{2}}{2}$
60°	$\dfrac{\pi}{3}$	$\dfrac{\sqrt{3}}{2}$	$\dfrac{1}{2}$
90°	$\dfrac{\pi}{2}$	1	0
180°	π	0	−1

4. Pythagorean identities: The following identities, based on the Pythagorean Theorem, will help you simplify complicated trigonometric expressions.

$$\sin^2 \theta + \cos^2 \theta = 1$$
$$\tan^2 \theta + 1 = \sec^2 \theta$$
$$\cot^2 \theta + 1 = \csc^2 \theta$$

5. Double-angle identities: These identities will help you simplify double-angle trigonometric expressions into single-angle expressions.

$$\sin 2\theta = 2 \sin \theta \cos \theta$$
$$\cos 2\theta = \cos^2 \theta - \sin^2 \theta$$
$$= 2 \cos^2 \theta - 1$$
$$= 1 - 2 \sin^2 \theta$$

STEPS YOU NEED TO REMEMBER

1. *Simplify the trigonometric expression.*

To solve trigonometric equations containing multiple trigonometric functions, you should first simplify the expressions using trigonometric identities. Identities are equations that are true for every value of x. There are many types of trigonometric identities, but the ones you should know by heart are the reciprocal and ratio identities, Pythagorean identities, and double-angle identities.

For example, $\sin\theta + \dfrac{\cos\theta}{\cot\theta} = \sqrt{3}$ simplifies to $2\sin\theta = \sqrt{3}$.

2. *Solve for the trigonometric function.*

Once you have simplified the expression, you can solve the equation for the trigonometric function.

$$2\sin\theta = \sqrt{3}$$
$$\sin\theta = \frac{\sqrt{3}}{2}$$

3. *Find the measure of the angle of the trigonometric functions.*

If you know the value of the trigonometric function, you can find the measure of the angle.

We know that $\sin\dfrac{\pi}{3} = \dfrac{\sqrt{3}}{2}$. Therefore, $\theta = \dfrac{\pi}{3}$.

4. *Give all possible measures of the angle on the given interval.*

When you are asked to solve a trigonometric equation, you will be given an interval such as $[0,2\pi]$. You must give all possible measures of the angle on the given interval that satisfies the equation.

The values for θ on the interval $[0,2\pi]$ that satisfy the equation $\sin\theta = \dfrac{\sqrt{3}}{2}$ are $\dfrac{\pi}{3}$ and $\dfrac{2\pi}{3}$.

COMMON TRIGONOMETRIC FUNCTION QUESTIONS

Sine and Cosine Values of Angles: Find the sine and cosine values for the following angles:

(a) $\dfrac{\pi}{6}$ (b) $\dfrac{5\pi}{6}$ (c) $\dfrac{7\pi}{6}$ (d) $\dfrac{11\pi}{6}$

Step 1: Draw the angle in standard position on an *xy* plane.

Step 2: Using the terminal side of the angle as the hypotenuse, draw a right triangle with a hypotenuse of 1.

Step 3: Find the lengths of the legs of the right triangle.

Step 4: Use the trigonometric ratios to find the values of the trigonometric functions.

> **Solution and Explanation:** You have been given a table of trigonometric values to memorize. Problem (a) illustrates how to arrive at those values mathematically. Problems (b), (c), and (d) will show how to find the values of angles that are not listed in the table.
>
> (a) Using the conversion ratio $\dfrac{180°}{\pi}$, we find that $\dfrac{\pi}{6}$ radians = 30°:
>
> $$\left(\dfrac{\pi}{6}\right)\left(\dfrac{180}{\pi}\right) = \dfrac{180}{6} = 30°$$
>
> Therefore, an angle of $\dfrac{\pi}{6}$ radians in standard form on an *xy* plane has its terminal side in quadrant I.

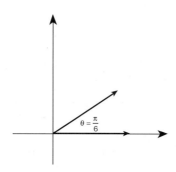

Next, we construct a right triangle with a hypotenuse of 1.

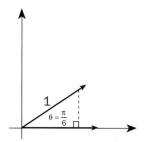

Because $\frac{\pi}{6} = 30°$, we know that the triangle is a 30-60-90 triangle and the sides of the triangle are in the ratio 1:2: $\sqrt{3}$.

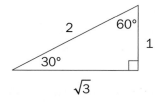

In our triangle, the length of the hypotenuse is 1, so the length of the side opposite the 30° angle is $\frac{1}{2}$ and the length of the side opposite the 60° angle is $\frac{\sqrt{3}}{2}$.

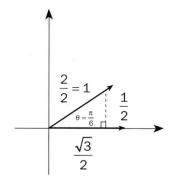

Now that we have the sides of the triangle, we can find the sine and cosine values of the angle:

$$\sin\frac{\pi}{6} = \frac{y}{r} = \frac{\frac{1}{2}}{1} = \frac{1}{2} \qquad \cos\frac{\pi}{6} = \frac{x}{r} = \frac{\frac{\sqrt{3}}{2}}{1} = \frac{\sqrt{3}}{2}$$

(b) Using the conversion ratio $\frac{180°}{\pi}$, we find that $\frac{5\pi}{6}$ radians = 150°, which lies in quadrant II.

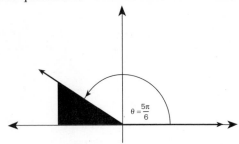

To construct a right triangle on the *xy* plane, we use the reference angle. The reference angle is the smallest angle between the terminal side and the *x*-axis. The reference angle for $\frac{5\pi}{6}$ is $\pi - \frac{5\pi}{6} = \frac{\pi}{6}$. (Recall that π radians = 180°.)

From problem (a), we know that the right triangle is a 30-60-90 triangle and the lengths of its legs are $\frac{\sqrt{3}}{2}$ and $\frac{1}{2}$. However, since the triangle is in quadrant II, we assign a negative value to the base because it lies along the negative *x*-axis.

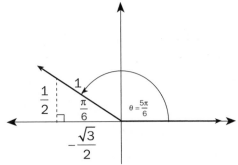

Therefore,

$$\sin \frac{5\pi}{6} = \frac{y}{r} = \frac{1}{2}$$

$$\cos \frac{5\pi}{6} = \frac{x}{r} = -\frac{\sqrt{3}}{2}$$

(c) The angle $\frac{7\pi}{6}$ radians lies in quadrant III and its reference angle is $\frac{7\pi}{6} - \pi = \frac{\pi}{6}$.

Once again, we have a 30-60-90 triangle, but now the values of both legs are negative because one lies along the negative x-axis and the other lies along the negative y-axis.

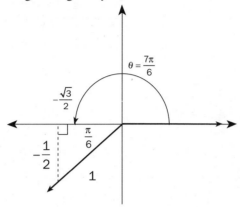

Therefore,

$$\sin\frac{7\pi}{6} = -\frac{1}{2}$$

$$\cos\frac{7\pi}{6} = -\frac{\sqrt{3}}{2}$$

(d) The angle $\frac{11\pi}{6}$ radians lies in quadrant IV and its reference angle is $2\pi - \frac{11\pi}{6} = \frac{\pi}{6}$. ($2\pi$ radians = 360°.)

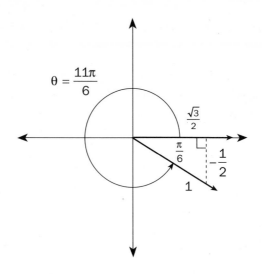

Therefore,

$$\sin \frac{11\pi}{6} = -\frac{1}{2}$$

$$\cos \frac{11\pi}{6} = \frac{\sqrt{3}}{2}$$

By now, you should have noticed a pattern: the trigonometric values of multiples of the same common angle repeat themselves. Hence, trigonometric functions are periodic functions. By memorizing the table of sine and cosine values for $\frac{\pi}{6}, \frac{\pi}{4}, \frac{\pi}{3}, \frac{\pi}{2}$, and π, you can quickly determine the values of trigonometric functions for multiples of those angles by the quadrant in which the angles lie.

But what if a trigonometric expression contains more than one trigonometric function, including tangent, cosecant, secant, and cotangent functions? The best strategy is to simplify these expressions.

Simplify Using Reciprocal and Ratio Identities: Simplify the following expressions using reciprocal and ratio identities:

(a) $\dfrac{\sin\theta}{\csc\theta}$ (b) $\sec\theta\cos\theta$ (c) $\sin\theta + \dfrac{\cos\theta}{\cot\theta}$

Step 1: Identify the trigonometric functions and their identities.

Step 2: Substitute the functions with the identities that will simplify the expression to a single trigonometric function.

Step 3: Simplify.

Solution and Explanation: The reciprocal and ratio identities are merely the relationships between the trigonometric functions:

Reciprocal Identities		Ratio Identities
$\sin\theta = \dfrac{1}{\csc\theta}$	$\csc\theta = \dfrac{1}{\sin\theta}$	$\tan\theta = \dfrac{\sin\theta}{\cos\theta}$
$\cos\theta = \dfrac{1}{\sec\theta}$	$\sec\theta = \dfrac{1}{\cos\theta}$	$\cot\theta = \dfrac{\cos\theta}{\sin\theta}$
$\tan\theta = \dfrac{1}{\cot\theta}$	$\cot\theta = \dfrac{1}{\tan\theta}$	

(a) To simplify $\dfrac{\sin\theta}{\csc\theta}$, we can use the identity $\sin\theta = \dfrac{1}{\csc\theta}$ or $\csc\theta = \dfrac{1}{\sin\theta}$.

If we use the identity $\sin\theta = \dfrac{1}{\csc\theta}$, we get $\dfrac{\sin\theta}{\csc\theta} = \dfrac{\frac{1}{\csc\theta}}{\csc\theta} = \dfrac{1}{\csc^2\theta}$.

We can simplify further using the identity $\csc\theta = \dfrac{1}{\sin\theta}$ to get

$$\dfrac{1}{\csc^2\theta} = \dfrac{1}{\frac{1}{\sin^2\theta}} = \sin^2\theta.$$

On the other hand, we could have used the identity $\csc\theta = \dfrac{1}{\sin\theta}$ and arrived at the same answer.

$$\dfrac{\sin\theta}{\csc\theta} = \dfrac{\sin\theta}{\frac{1}{\sin^2\theta}} = \sin^2\theta$$

(b) To simplify $\sec\theta\cos\theta$, we can use the identity $\sec\theta = \dfrac{1}{\cos\theta}$ or $\cos\theta = \dfrac{1}{\sec\theta}$.

$$\sec\theta\cos\theta = \frac{1}{\cos\theta}\cdot\cos\theta = 1 \quad \text{or} \quad \sec\theta\cos\theta = \sec\theta\cdot\frac{1}{\sec\theta} = 1$$

(c) To simplify $\sin\theta + \dfrac{\cos\theta}{\cot\theta}$, we can use the identity $\cot\theta = \dfrac{1}{\tan\theta}$ or $\cot\theta = \dfrac{\sin\theta}{\cos\theta}$.

$$\sin\theta + \frac{\cos\theta}{\cot\theta} = \sin\theta + \frac{\cos\theta}{\dfrac{1}{\tan\theta}}$$
$$= \sin\theta + \tan\theta\cos\theta$$

Next, we use the identity $\tan\theta = \dfrac{\sin\theta}{\cos\theta}$ to get

$$\sin\theta + \frac{\sin\theta}{\cos\theta}\cdot\cos\theta = \sin\theta + \sin\theta$$
$$= 2\sin\theta$$

If we had used the identity $\cot\theta = \dfrac{\cos\theta}{\sin\theta}$, we would have arrived at the same answer in fewer steps:

$$\sin\theta + \frac{\cos\theta}{\dfrac{\cos\theta}{\sin\theta}} = \sin\theta + \frac{\sin\theta}{\cos\theta}\cdot\cos\theta$$
$$= \sin\theta + \sin\theta$$
$$= 2\sin\theta$$

For more complicated expressions containing trigonometric functions that are squared, such as $\sin^2\theta$, you will need to use the Pythagorean identities.

Simplify Using Pythagorean Identities: Simplify the following trigonometric expressions using the Pythagorean identities:

(a) $\sin\theta\,(1 + \cot^2\theta)$ (b) $\sec^2\theta - \tan\theta\cot\theta$ (c) $\cos\theta + \tan\theta\sin\theta$

Step 1: Wherever possible, simplify using the reciprocal and ratio identities.

Step 2: Identify the squared trigonometric function and its Pythagorean identity.

Step 3: Substitute the trigonometric function with its equivalent expression.

Step 4: Simplify.

Solution and Explanation: The Pythagorean identities are based on the Pythagorean Theorem, $a^2 + b^2 = c^2$:

$$\sin^2\theta + \cos^2\theta = 1$$
$$\tan^2\theta + 1 = \sec^2\theta$$
$$\cot^2\theta + 1 = \csc^2\theta$$

(a) To simplify the expression $\sin\theta\,(1 + \cot^2\theta)$, we use the Pythagorean identity

$$\cot^2\theta + 1 = \csc^2\theta.$$

$$\sin\theta\,(1 + \cot^2\theta) = \sin\theta\,(\csc^2\theta)$$

We know $\csc\theta = \dfrac{1}{\sin\theta}$, so we can simplify even further.

$$\sin\theta\left(\csc^2\theta\right) = \sin\theta\left(\frac{1}{\sin^2\theta}\right)$$

$$= \frac{1}{\sin\theta}$$

Because $\csc\theta = \dfrac{1}{\sin\theta}$, our final answer is $\csc\theta$.

(b) To simplify the expression $\sec^2\theta - \tan\theta\cot\theta$, we can first use the identity $\cot\theta = \dfrac{1}{\tan\theta}$:

$$\sec^2\theta - \tan\theta\cot\theta = \sec^2\theta - \tan\theta \cdot \frac{1}{\tan\theta}$$

$$= \sec^2\theta - 1$$

Next, we use the Pythagorean identity $\tan^2\theta + 1 = \sec^2\theta$.

$$\sec^2\theta - 1 = \left(\tan^2\theta + 1\right) - 1$$

$$= \tan^2\theta$$

(c) To simplify $\cos\theta + \tan\theta\sin\theta$, we first use the identity $\tan\theta = \dfrac{\sin\theta}{\cos\theta}$.

$$\cos\theta + \tan\theta\sin\theta = \cos\theta + \frac{\sin\theta}{\cos\theta}\cdot\sin\theta$$

$$= \cos\theta + \frac{\sin^2\theta}{\cos\theta}$$

To add $\cos\theta + \dfrac{\sin^2\theta}{\cos\theta}$, we need a common denominator of $\cos\theta$:

$$\cos\theta + \frac{\sin^2\theta}{\cos\theta} = \frac{\cos^2\theta}{\cos\theta} + \frac{\sin^2\theta}{\cos\theta}$$

$$= \frac{\cos^2\theta + \sin^2\theta}{\cos\theta}$$

Now we have an expression in the numerator that is equivalent to 1, according to the Pythogorean identity $\cos^2\theta + \sin^2\theta = 1$.

$$\frac{\cos^2\theta + \sin^2\theta}{\cos\theta} = \frac{1}{\cos\theta} = \sec\theta$$

If an expression contains double-angles, you will need to use the double-angle identities.

Double-Angle Identities: Simplify using the double-angle identities:

(a) $\dfrac{\sin 2\theta}{\tan\theta}$ (b) $\cos 2\theta - \cos^2\theta$ (c) $\dfrac{1 - 2\sin^2\theta}{\sin 2\theta}$

Step 1: Wherever possible, simplify using the reciprocal and ratio identities.

Step 2: Identify the double-angle expressions and their identities.

Step 3: Substitute the double-angle expressions with their equivalent single-angle expressions.

Step 4: Simplify.

Solution and Explanation: The double-angle identities are:

$$\sin 2\theta = 2\sin\theta\cos\theta$$

$$\cos 2\theta = \cos^2\theta - \sin^2\theta$$

$$= 2\cos^2\theta - 1$$

$$= 1 - 2\sin^2\theta$$

(a) To simplify $\dfrac{\sin 2\theta}{\tan \theta}$, we can first use the identity $\tan\theta = \dfrac{\sin\theta}{\cos\theta}$:

$$\frac{\sin 2\theta}{\tan\theta} = \frac{\sin 2\theta}{\dfrac{\sin\theta}{\cos\theta}} = \frac{\cos\theta \sin 2\theta}{\sin\theta}$$

Next, we use the double-angle identity $\sin 2\theta = 2\sin\theta\cos\theta$.

$$\frac{\cos\theta \cdot 2\sin\theta\cos\theta}{\sin\theta} = \frac{\cos^2\theta \cdot 2\,\cancel{\sin\theta}}{\cancel{\sin\theta}} = 2\cos^2\theta$$

(b) To simplify $\cos 2\theta - \cos^2\theta$, we can use the identity $\cos 2\theta = \cos^2\theta - \sin^2\theta$:

$$\cos 2\theta - \cos^2\theta = (\cos^2\theta - \sin^2\theta) - \cos^2\theta = -\sin^2\theta$$

We could also have used the identity $\cos 2\theta = 2\cos^2\theta - 1$.

$$\cos 2\theta - \cos^2\theta = (2\cos^2\theta - 1) - \cos^2\theta = \cos^2\theta - 1.$$

The Pythagorean identity $\sin^2\theta + \cos^2\theta = 1$ gives us $\cos^2\theta - 1 = -\sin^2\theta$.

(c) Simplifying $\dfrac{1 - 2\sin^2\theta}{\sin 2\theta}$ is a bit trickier. If we use the identity

$\sin 2\theta = 2\sin\theta\cos\theta$ we get $\dfrac{1 - 2\sin^2\theta}{\sin 2\theta} = \dfrac{1 - 2\sin^2\theta}{2\sin\theta\cos\theta}$, which turns out to be a more complicated expression than the original one.

However, using the identity $\cos 2\theta = 1 - 2\sin^2\theta$ give us

$$\frac{1 - 2\sin^2\theta}{\sin 2\theta} = \frac{\cos 2\theta}{\sin 2\theta}.$$

Because $\cot\theta = \dfrac{\cos\theta}{\sin\theta}$, $\dfrac{\cos 2\theta}{\sin 2\theta} = \cot 2\theta$.

Now that you know how to simplify trigonometric expressions, you can solve trigonometric equations.

Trigonometric Equations: Find all solutions of the equation $\cos 2\theta - \cos\theta = 0$ on the interval $[0, 2\pi]$.

Step 1: Simplify the expression using the identities.

Step 2: Solve for the trigonometric function.

Step 3: Find the measures of the angles on the given interval that satisfy the value of the trigonometric function.

Solution and Explanation: To simplify the expression, review the identities from the previous problems. There are many more identities, but the ones we have covered here are the ones you should memorize.

We will first use one of the double-angle formulas to replace $\cos 2\theta$ with its equivalent single-angle expression.

There are three possibilities for the single-angle expression:

$$\left(\cos^2 \theta - \sin^2 \theta\right) - \cos\theta = 0$$
$$\left(2\cos^2 \theta - 1\right) - \cos\theta = 0$$
$$\left(1 - 2\sin^2 \theta\right) - \cos\theta = 0$$

Of the three possibilities, the second one is simplest because it contains only cosine terms, so we will use it.

By rearranging the terms in $(2\cos^2 \theta - 1) - \cos\theta = 0$, we get $2\cos^2 \theta - \cos\theta - 1 = 0$, which looks very much like a quadratic equation $ax^2 + bx + c = 0$. Remember that you can factor quadratic expressions.

Therefore,

$$2\cos^2 \theta - \cos\theta - 1 = 0$$
$$\left(2\cos\theta + 1\right)\left(\cos\theta - 1\right) = 0$$

Set each factor equal to zero and solve for $\cos\theta$.

$$2\cos\theta + 1 = 0$$
$$2\cos\theta = -1 \quad \text{and} \quad \begin{aligned} \cos\theta - 1 &= 0 \\ \cos\theta &= 1 \end{aligned}$$
$$\cos\theta = -\frac{1}{2}$$

$\cos\theta = -\frac{1}{2}$ when the reference angle $\theta = \frac{\pi}{3}$ is in quadrants II and III.

In quadrant II, $\cos\dfrac{2\pi}{3} = -\dfrac{1}{2}$.

In quadrant III, $\cos\dfrac{4\pi}{3} = -\dfrac{1}{2}$.

$\cos\theta = 1$ when $\theta = 0$ radians and 2π radians.

Therefore, all the possible values for θ on the interval $[0, 2\pi]$ are $\theta = 0$, $\dfrac{2\pi}{3}$, $\dfrac{4\pi}{3}$, and 2π.

CHAPTER QUIZ

1. Find the value of $\cos\dfrac{9\pi}{4}$.

   (A) $\dfrac{\sqrt{2}}{2}$

   (B) $-\dfrac{\sqrt{2}}{2}$

   (C) $\dfrac{\sqrt{3}}{2}$

   (D) $-\dfrac{\sqrt{3}}{2}$

   (E) 1

2. Find the value of $\tan\dfrac{3\pi}{4}$.

   (A) 0

   (B) 1

   (C) -1

   (D) $\dfrac{1}{2}$

   (E) $-\dfrac{1}{2}$

3. Simplify $\dfrac{\cos^2\theta}{\sin\theta} + \sin\theta$

   (A) $\sin\theta$

   (B) $\cos\theta$

   (C) $\tan\theta$

   (D) $\cot\theta$

   (E) $\csc\theta$

4. Simplify $\dfrac{\sin\theta}{\csc\theta} + \dfrac{\cos\theta}{\sec\theta}$

   (A) 0

   (B) 1

   (C) $\sin^2\theta$

   (D) $\cos^2\theta$

   (E) $\sec\theta$

5. Simplify $\dfrac{\cos^2\theta - \sin^2\theta}{\sin 2\theta}$

   (A) $\sin 2\theta$

   (B) $\sec 2\theta$

   (C) $\tan 2\theta$

   (D) $\cot 2\theta$

   (E) $\csc 2\theta$

6. Simplify $2\sin\theta\cos\theta - 4\sin^3\theta\cos\theta$

   (A) $\sin 2\theta$

   (B) $\cos 2\theta$

   (C) $\sin 2\theta\cos 2\theta$

   (D) $\tan 2\theta$

   (E) $\sec 2\theta\csc 2\theta$

7. Find the value of $\sin 2\theta$ on the interval $\left[\pi, \dfrac{3\pi}{2}\right]$ if $\cos\theta = -\dfrac{3}{5}$.

   (A) $\dfrac{24}{25}$

   (B) $\dfrac{3}{5}$

   (C) $\dfrac{4}{5}$

   (D) $\dfrac{\sqrt{3}}{2}$

   (E) $\dfrac{1}{2}$

8. If θ is in quadrant II and $\sin\theta = \dfrac{5}{13}$, what is $\cos 2\theta$?

 (A) $-\dfrac{12}{13}$

 (B) $-\dfrac{5}{12}$

 (C) $-\dfrac{144}{169}$

 (D) $-\dfrac{119}{169}$

 (E) $\dfrac{119}{169}$

9. Solve $4(1 - \cos^2\theta) = 1$ on the interval $[0,2\pi]$.

 (A) $\dfrac{\pi}{6}$

 (B) $\dfrac{5\pi}{6}$

 (C) $\dfrac{7\pi}{6}$

 (D) $\dfrac{11\pi}{6}$

 (E) All of the above.

10. Solve $\sin\theta\cos\theta = -\sin\theta$ on the interval $\left(0, \dfrac{3\pi}{2}\right)$.

 (A) 0

 (B) π

 (C) $\dfrac{\pi}{2}$

 (D) $\dfrac{\pi}{4}$

 (E) None of the above.

Answers and Explanations

1. A

The angle $\frac{9\pi}{4}$ radians is greater than 2π radians, which is the measure of a complete circle.

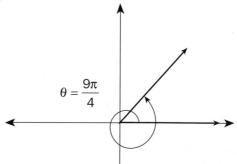

To find the reference angle, we subtract $\frac{9\pi}{4} - 2\pi = \frac{9\pi}{4} - \frac{8\pi}{4} = \frac{\pi}{4}$.

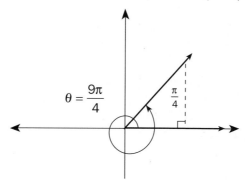

From the graph above, we see that $\cos\frac{9\pi}{4} = \cos\frac{\pi}{4} = \frac{\sqrt{2}}{2}$.

2. C

We use our knowledge of the sine and cosine values of $\frac{3\pi}{4}$ as well as the ratio identity $\tan\theta = \frac{\sin\theta}{\cos\theta}$ to solve this problem.

$$\tan\frac{3\pi}{4} = \frac{\sin\frac{3\pi}{4}}{\cos\frac{3\pi}{4}} = \frac{\frac{\sqrt{2}}{2}}{-\frac{\sqrt{2}}{2}} = -1$$

3. E

To simplify $\dfrac{\cos^2 \theta}{\sin \theta} + \sin \theta$, we first add the two terms using the common denominator $\sin\theta$.

$$\frac{\cos^2 \theta}{\sin \theta} + \sin \theta = \frac{\cos^2 \theta}{\sin \theta} + \frac{\sin^2 \theta}{\sin \theta}$$

$$= \frac{\cos^2 \theta + \sin^2 \theta}{\sin \theta}$$

Next, we use the Pythagorean identity $\sin^2\theta + \cos^2\theta = 1$ followed by $\csc\theta = \dfrac{1}{\sin\theta}$.

$$\frac{\cos^2 \theta + \sin^2 \theta}{\sin \theta} = \frac{1}{\sin \theta} = \csc \theta.$$

4. B

To simplify $\dfrac{\sin \theta}{\csc \theta} + \dfrac{\cos \theta}{\sec \theta}$, we start off with $\csc \theta = \dfrac{1}{\sin \theta}$ and $\sec \theta = \dfrac{1}{\cos \theta}$.

$$\frac{\sin \theta}{\csc \theta} + \frac{\cos \theta}{\sec \theta} = \frac{\sin \theta}{\dfrac{1}{\sin \theta}} + \frac{\cos \theta}{\dfrac{1}{\cos \theta}} = \sin^2 \theta + \cos^2 \theta$$

Because $\sin^2\theta + \cos^2\theta = 1$, our answer is 1.

5. D

To simplify $\dfrac{\cos^2 \theta - \sin^2 \theta}{\sin 2\theta}$ in the fewest number of steps, we need to

recognize that $\cos^2\theta - \sin^2\theta = \cos2\theta$.

Therefore, $\dfrac{\cos^2 \theta - \sin^2 \theta}{\sin 2\theta} = \dfrac{\cos 2\theta}{\sin 2\theta} = \cot 2\theta.$

6. C

The expression $2\sin\theta \cos\theta - 4 \sin^3\theta \cos\theta$ does not contain any double-angles, but we can begin by factoring out the expression $2\sin\theta \cos\theta$.

$$2\sin\theta \cos\theta - 4 \sin^3\theta \cos\theta = 2\sin\theta \cos\theta (1 - 2\sin^2\theta)$$

This time we work backwards with the double-angle identities. We simplify by converting single-angle expressions to their double-angle equivalents.

$$2\sin\theta \cos\theta = \sin2\theta$$

$$1 - 2\sin^2\theta = \cos2\theta$$

Therefore, $2\sin\theta \cos\theta (1 - 2\sin^2\theta) = \sin2\theta \cos2\theta.$

7. A

To use the double-angle identity $\sin 2\theta = 2\sin\theta\cos\theta$, we need to first find $\sin\theta$. To do so, we will use the Pythagorean identity $\sin^2\theta + \cos^2\theta = 1$.

$$\sin^2\theta + \cos^2\theta = 1$$
$$\sin^2\theta = 1 - \cos^2\theta$$
$$= 1 - \left(-\frac{3}{5}\right)^2$$
$$= \frac{16}{25}$$
$$\sin\theta = \sqrt{\frac{16}{25}} = \pm\frac{4}{5}$$

Since $\sin\theta$ on the interval $\left[\pi, \frac{3\pi}{2}\right]$ is in quadrant III, $\sin\theta = -\frac{4}{5}$.

Now we can find $\sin 2\theta$:
$$\sin 2\theta = 2\sin\theta\cos\theta$$
$$= 2\left(-\frac{4}{5}\right)\left(-\frac{3}{5}\right)$$
$$= \frac{24}{25}$$

8. D

We use the identity $\cos 2\theta = 1 - 2\sin^2\theta$.

$$\cos 2\theta = 1 - 2\sin^2\theta$$
$$= 1 - 2\left(\frac{5}{13}\right)^2$$
$$= \frac{119}{169}$$

Since the cosine function is negative in quadrant II, $\cos 2\theta = -\frac{119}{169}$.

9. E

We use the identity $\sin^2\theta = (1-\cos^2\theta)$ to simplify $4(1-\cos^2\theta)$.

$$4\left(1-\cos^2\theta\right)=1$$
$$4\left(\sin^2\theta\right)=1$$
$$\sin^2\theta=\frac{1}{4}$$
$$\sin\theta=\sqrt{\frac{1}{4}}=\pm\frac{1}{2}$$

On the interval $[0,2\pi]$, $\theta=\dfrac{\pi}{6},\dfrac{5\pi}{6},\dfrac{7\pi}{6},\dfrac{11\pi}{6}$.

10. B

Recall from Chapter 3 that we solve an equation by setting it equal to zero. We can do the same with $\sin\theta\cos\theta=-\sin\theta$.

$$\sin\theta\cos\theta=-\sin\theta$$
$$\sin\theta\cos\theta+\sin\theta=0$$

We can factor out $\sin\theta$ and then set the factors equal to zero.

$$\sin\theta\cos\theta+\sin\theta=0 \qquad\qquad \cos\theta+1=0$$
$$\sin\theta\left(\cos\theta+1\right)=0 \qquad\qquad \cos\theta=-1$$
$$\sin\theta=0 \qquad\qquad\qquad\qquad \theta=\pi$$
$$\qquad\qquad\qquad\text{or}$$
$$\theta=0,\pi$$

Since our interval $\left(0,\dfrac{3\pi}{2}\right)$ does not include 0, we are left with $\theta=\pi$.

CHAPTER 6

Limits and Continuity

WHAT ARE LIMITS AND CONTINUITY?

Limits are a way of describing how a function behaves *near* a point (but not at a point). We use the notation $\lim\limits_{x \to c} f(x) = L$ to indicate that as x gets close to a number c, $f(x)$ gets close to a number L. We read $\lim\limits_{x \to c} f(x) = L$ as "the limit of $f(x)$ as x approaches c is L."

The continuity of a function refers to the way a function behaves *at* a point or points on an interval. If there are no holes or breaks in the function at $x = c$, the function is said to be continuous at $x = c$.

In this chapter, we will also explore instances in which a function *does not* approach a finite value L and is *discontinuous*.

CONCEPTS TO HELP YOU

1. One-sided limits: If $f(x)$ approaches a number L as x approaches c from the *left*, we say that the left-hand limit exists and is equal to L and we write

    $$\lim_{x \to c^-} f(x) = L.$$

    If $f(x)$ approaches a number L as x approaches c from the *right*, we say that the right-hand limit exists and is equal to L and we write $\lim\limits_{x \to c^+} f(x) = L$.

The limit of a function exists at a point $x = c$ if—and only if—the left-hand limit and the right-hand limit both exist and are equal:

If $\lim\limits_{x \to c^-} f(x) = L$

and $\lim\limits_{x \to c^+} f(x) = L$

then $\lim\limits_{x \to c} f(x) = L$.

2. Operations with limits: If $\lim\limits_{x \to c} f(x) = L$ and $\lim\limits_{x \to c} g(x) = M$, then

	Rule (formula)	Rule in words
1)	$\lim\limits_{x \to c} [f(x) \pm g(x)] = \lim\limits_{x \to c} f(x) \pm \lim\limits_{x \to c} g(x) = L \pm M$	The limit of the sum (or difference) is the sum (or difference) of the limits
2)	$\lim\limits_{x \to c} [f(x) \cdot g(x)] = \left(\lim\limits_{x \to c} f(x) \right)\left(\lim\limits_{x \to c} g(x) \right) = L \cdot M$	The limit of the product is the product of the limits
3)	$\lim\limits_{x \to c} \dfrac{f(x)}{g(x)} = \dfrac{\lim\limits_{x \to c} f(x)}{\lim\limits_{x \to c} g(x)} = \dfrac{L}{M}$, for $M \neq 0$	The limit of the quotients is the quotient of the limits
4)	$\lim\limits_{x \to c} \sqrt{f(x)} = \sqrt{\lim\limits_{x \to c} f(x)} = \sqrt{L}$, (for L > 0, if n is even)	The limit of the roots is the root of the limit

3. Special trigonometric limits:

$$\lim_{x \to 0} \frac{\sin x}{x} = 1 \text{ and } \lim_{x \to 0} \frac{\cos x - 1}{x} = 0$$

4. Continuity: A function $f(x)$ is continuous at the point $x = c$ if all three conditions are met:

(1) $f(x)$ is defined at $x = c$

(2) $\lim\limits_{x \to c} f(x)$ exists

(3) $\lim\limits_{x \to c} f(x) = f(c)$

STEPS YOU NEED TO REMEMBER

1. *Substitute.*

When we are asked to compute $\lim\limits_{x \to c} f(x)$, the first thing we do is try to plug the value c into the function. If we get an answer that is a real number, we are done.

2. *Factor and conjugate.*

If we plug c in and get the indeterminate form $\frac{0}{0}$, we need to simplify the function. We can then plug c into the simplified function.

There are two general simplification strategies that will get us through many problems: factoring and conjugating. For a review of factoring, refer to Chapter 3. The conjugate of an expression of the form $a + b$ is the expression $a - b$. Sometimes multiplying an expression by its conjugate will greatly simplify the function. Limits that have sums or differences involving square roots are good candidates for conjugation.

3. *Create a table of values.*

If substitution, factoring, and conjugation do not work, evaluate numerically by creating a table of values for x and $f(x)$.

4. *Evaluate continuity.*

A function $f(x)$ is continuous at the point $x = c$ if

   (1) $f(x)$ is defined at $x = c$

   (2) $\lim\limits_{x \to c} f(x)$ exists

   (3) $\lim\limits_{x \to c} f(x) = f(c)$

5. *All polynomials are continuous.*

COMMON LIMITS AND CONTINUITY QUESTIONS

Evaluate Limits and Continuity Graphically: Evaluate the limits and continuity of the following graphs as x approaches 1:

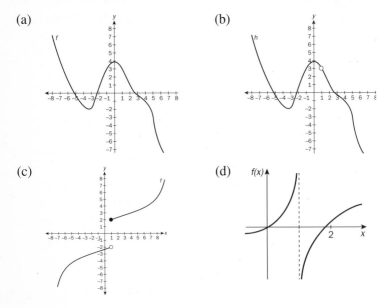

(a)

(b)

(c)

(d)

Step 1: Follow the curve as x approaches 1 from the left and from the right to evaluate the limit:

- If both sides of the curve meet at $f(1)$, then the limit is $f(1)$.
- If the left and right sides of the curve do not meet, then the limit does not exist and the function is discontinuous at $x = 1$.
- If the left and right sides of the curve do not meet and there is a vertical asymptote at $x = 1$, the limit is infinity and the function is discontinuous at $x = 1$.

Step 2: If the limit exists, locate $f(1)$ on the graph:

- If the limit exists and you can trace the curve as it passes through $(1, f(1))$ without lifting your pencil, the function is continuous at $x = 1$.
- If the limit exists and there is a hole in the function at $x = 1$, the function is discontinuous at $x = 1$.
- If the limit exists but is not the same as $f(1)$, the function is discontinuous at $x = 1$.

Solution and Explanation:

(a) As x approaches 1 from the left-hand side (that is, from $x = -8$ to $x = 1$), the graph approaches $f(x) = 3$. As x approaches 1 from the right-hand side (that is, from $x = 8$ to $x = 1$) the graph also approaches $f(x) = 3$. At $x = 1$, $f(x) = 3$.

We say, then, that the limit of $f(x)$ as x approaches 1 from the *left* is 3, and we write $\lim\limits_{x \to 1^-} f(x) = 3$.

The limit of $f(x)$ as x approaches 1 from the *right* is 3, and we write $\lim\limits_{x \to 1^+} f(x) = 3$.

Because the left- and right-hand limits are equal, the limit of $f(x)$ as x approaches 1 is 3 and the limit notation is $\lim\limits_{x \to 1} f(x) = 3$.

There are no holes or breaks in the graph at $x = 1$. Therefore, the function is *continuous* at $x = 1$. In fact, there are no holes or breaks anywhere on the graph, so the function is continuous everywhere.

We can also think about continuity in this way: A function is continuous if you can draw it without lifting your pencil.

(b) This graph is similar to the previous graph in that both the left- and right-hand limits approach 3 as x approaches 1. Therefore, the limit of $f(x)$ as x approaches 1 is 3.

$$\lim\limits_{x \to 1^-} f(x) = 3$$

$$\lim\limits_{x \to 1^+} f(x) = 3$$

$$\lim\limits_{x \to 1} f(x) = 3$$

However, there is a hole or circle on the graph where $x = 1$. This means that $f(1) \neq 3$. In fact, $f(1)$ is not equal to any number; it is undefined. The function is said to be *discontinuous* at $x = 1$. This type of discontinuity is called a *point discontinuity*.

The function is continuous everywhere else because there are no other holes.

(c) The solid circle indicates that $f(1) = 2$.

But as x approaches 1 from the left, $f(x)$ approaches -2: $\lim\limits_{x \to 1^-} f(x) = -2$.

As x approaches 1 from the right, $f(x)$ approaches 2: $\lim\limits_{x \to 1^+} f(x) = 2$.

Because $\lim\limits_{x \to 1^-} f(x) \neq \lim\limits_{x \to 1^+} f(x)$, the limit at $x = 1$ does not exist, and we write $\lim\limits_{x \to 1} f(x)$ DNE.

Because there is a break in the function at $x = 1$, the function is said to have a *jump discontinuity* at $x = 1$.

(d) The dotted vertical line at $x = 1$ is called a *vertical asymptote*. The asymptote indicates that as x approaches 1, $f(x)$ increases or decreases without bound and $f(1)$ is undefined. When this occurs, the function is said to have an *infinite limit* and an *infinite discontinuity*. The symbol for infinity is ∞.

As x approaches 1 from the left, $f(x)$ *increases* without bound:

$$\lim\limits_{x \to 1^-} f(x) = \infty.$$

As x approaches 1 from the right, $f(x)$ *decreases* without bound:

$$\lim\limits_{x \to 1^+} f(x) = -\infty$$

Because $\lim\limits_{x \to 1^-} f(x) \neq \lim\limits_{x \to 1^+} f(x)$, $\lim\limits_{x \to 1} f(x)$ DNE.

We have just evaluated limits and continuity graphically. We will now evaluate limits and continuity algebraically.

Evaluate Limits and Continuity Algebraically: Evaluate the limits and continuity algebraically:

(a) $\lim\limits_{x\to 3} x^2$ (b) $\lim\limits_{x\to 1} \dfrac{x^2-1}{x-1}$ (c) $\lim\limits_{x\to 4} \dfrac{4(x-4)}{\sqrt{x}-2}$

Step 1: Plug the value of c into $f(x)$. If $f(c)$ is a real number, the limit is $f(c)$.

Step 2: If $f(c)$ is the indeterminate form $\dfrac{0}{0}$, simplify the function by factoring or conjugating. Evaluate the limit by substituting c into the simplified function.

Step 3: Determine the continuity of the function by applying the definition of continuity. A function is continuous at $x = c$ if it meets the following conditions:

(1) $f(x)$ is defined at $x = c$

(2) $\lim\limits_{x\to c} f(x)$

(3) $\lim\limits_{x\to c} f(x) = f(c)$

Solution and Explanation:

(a) To solve $\lim\limits_{x\to 3} x^2$, we plug the number 3 into x^2 to get 9; therefore, $\lim\limits_{x\to 3} x^2 = 9$.

Furthermore, the function is continuous at $x = 3$ because

(1) $f(3) = 9$

(2) $\lim\limits_{x\to 3} x^2 = 9$

(3) $\lim\limits_{x\to 3} x^2 = f(3)$

(b) When we plug $x = 1$ into the function, we get $\frac{x^2-1}{x-1} = \frac{(1)^2-1}{1-1} = \frac{0}{0}$.
We call $\frac{0}{0}$ an "indeterminate form" because we don't get any information from this expression. To evaluate the limit, we need to simplify the problem by factoring.

$$\frac{x^2-1}{x-1} = \frac{(x+1)\cancel{(x-1)}}{\cancel{x-1}} = x+1$$

As long as we stay away from the actual value $x = 1$, the function $\frac{x^2-1}{x-1}$ and $x = 1$ are identical, and we can find the limit as x approaches 1.

$$\lim_{x\to 1} \frac{x^2-1}{x-1} = \lim_{x\to 1}(x+1) = 2$$

We say that a function has a point discontinuity at $x = c$ if $\lim_{x\to c} f(x)$ exists but $\lim_{x\to c} f(x) \neq f(c)$.

Therefore, the function $\frac{x^2-1}{x-1}$ has a point discontinuity at $x = 1$ because

(1) $f(1)$ is undefined

(2) $\lim_{x\to 1} \frac{x^2-1}{x-1} = 2$

(3) $\lim_{x\to 1} \frac{x^2-1}{x-1} \neq f(1)$

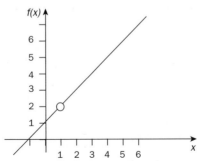

(c) To evaluate $\lim_{x\to 4} \frac{4(x-4)}{\sqrt{x}-2}$, we first try to plug in $x = 4$ and get the indeterminate form $\frac{0}{0}$. Because the denominator contains a radical, we can conjugate.

We multiply both the numerator and the denominator by the conjugate of the denominator.

$$\lim_{x \to 4} \frac{4(x-4)}{\sqrt{x}-2} = \lim_{x \to 4} \frac{4(x-4)}{\sqrt{x}-2} \cdot \frac{\sqrt{x}+2}{\sqrt{x}+2} = \lim_{x \to 4} \frac{4(x-4)(\sqrt{x}+2)}{x-4} = \lim_{x \to 4} \left(4\sqrt{x}+8\right)$$

After conjugating and simplifying, we are left with a limit we can evaluate by plugging in.

$$\lim_{x \to 4} \left(4\sqrt{x}+8\right) = 4\sqrt{4}+8 = 16$$

The function $\dfrac{4(x-4)}{\sqrt{x}-2}$ has a point discontinuity at $x = 4$ because

(1) $f(4)$ is undefined

(2) $\lim\limits_{x \to 4} \dfrac{4(x-4)}{\sqrt{x}-2} = 16$

(3) $\lim\limits_{x \to 4} \dfrac{4(x-4)}{\sqrt{x}-2} \neq f(4)$

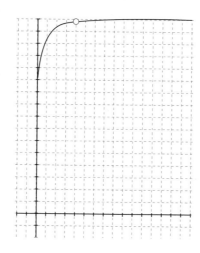

Sometimes, you will not be able to find a real number value for $f(x)$ because $f(x)$ increases or decreases without bound as x approaches c. We call these infinite limits.

If $f(x)$ increases without bound as x approaches c, the limit of $f(x)$ is positive infinity. If $f(x)$ decreases without bound as x approaches c, the limit of $f(x)$ is negative infinity.

Infinite Limits: Evaluate the limits and vertical asymptotes:

(a) $\lim\limits_{x\to 0} \dfrac{1}{x}$ (b) $\lim\limits_{x\to 0} \dfrac{1}{x^2}$ (c) $\lim\limits_{x\to\sqrt{2}} = \dfrac{x-1}{x^2-2}$

Step 1: Try substitution, factoring, and conjugation first.

Step 2: If substitution yields a fraction of the form $\dfrac{\text{something}}{0}$ and factorization and conjugation do not work, evaluate numerically by making a table of values for x and $f(x)$.

Step 3: If the left- and right-hand limits as x approaches c are both infinity, then there is a vertical asymptote at $x = c$.

> **Solution and Explanation:** When we try to substitute into a limit and get $\dfrac{\text{something}}{0}$, we are dealing with infinity. The limit will either be infinity, negative infinity, or will not exist (because the one-sided limits are positive infinity and negative infinity).
>
> When the limit is infinite (or both one-sided limits are infinite) at a point $x = c$, the function has a vertical asymptote at $x = c$. Because vertical asymptotes occur where the function increases without bound, points where the function is undefined are good places to check for vertical asymptotes.
>
> (a) If we try to solve $\lim\limits_{x\to 0} \dfrac{1}{x}$ by substitution, we get $\dfrac{1}{0}$. We cannot factor or conjugate this function. We can, however, evaluate the limit numerically by plugging in values to the left and right of 0.
>
> When we approach zero from the left, the value of $\dfrac{1}{x}$ gets smaller and smaller:

x	-0.1	-0.01	-0.001	-0.0001
$\dfrac{1}{x}$	-10	-100	-1000	$-10,000$

Therefore, $\lim\limits_{x\to 0^-} \dfrac{1}{x} = -\infty$.

When we approach zero from the right, the value of $\dfrac{1}{x}$ gets bigger and bigger:

x	0.1	0.01	0.001	0.0001
$\dfrac{1}{x}$	10	100	1000	$10,000$

Therefore, $\lim\limits_{x \to 0^+} \dfrac{1}{x} = \infty$

Because $\lim\limits_{x \to 0^-} \dfrac{1}{x} \neq \lim\limits_{x \to 0^+} \dfrac{1}{x}$, $\lim\limits_{x \to 0} \dfrac{1}{x}$ does not exist. We write this as

$\lim\limits_{x \to 0} \dfrac{1}{x} = \text{DNE}$.

However, a vertical asymptote exists at $x = 0$ because the function is undefined at $x = 0$ *and* the one-sided limits are both infinite.

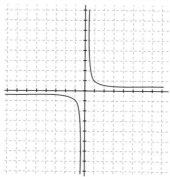

Another way to evaluate $\lim\limits_{x \to 0} \dfrac{1}{x}$ is to analyze the signs of the numerator and the denominator. As $x \to 0^-$, the numerator is fixed and positive, but the denominator is negative. The denominator is approaching zero, so the left-hand limit is negative infinity. As $x \to 0^+$, the numerator is fixed and positive, and the denominator is also positive. The denominator is approaching zero, so the right-hand limit is positive infinity. Because the one-sided limits are not equal, the limit does not exist.

(b) When we substitute 0 into $\dfrac{1}{x^2}$, we get $\dfrac{1}{0}$, or "$\dfrac{\text{something}}{0}$." The next step is to evaluate the left- and right-hand limits.

As $x \to 0^-$, the numerator is fixed and positive, and the denominator is a square so it is also positive. The denominator is approaching zero, so the left-hand limit is positive infinity. As $x \to 0^+$, the numerator is fixed and positive, and the denominator is also positive. The denominator is approaching zero, so the right-hand limit is positive infinity.

We can verify this by making a table of values. As we approach zero from the left, the values of $\dfrac{1}{x^2}$ get bigger and bigger:

x	−0.1	−0.01	−0.001	−0.0001
$\dfrac{1}{x^2}$	100	10,000	1,000,000	100,000,000

Therefore, $\lim\limits_{x \to 0^-} \dfrac{1}{x^2} = \infty.$

As we approach zero from the right, the values of $\dfrac{1}{x^2}$ also get bigger and bigger:

x	0.1	0.01	0.001	0.0001
$\dfrac{1}{x^2}$	100	10,000	1,000,000	100,000,000

Therefore, $\lim\limits_{x \to 0^+} \dfrac{1}{x^2} = \infty.$

Because $\lim\limits_{x \to 0^-} \dfrac{1}{x^2} = \lim\limits_{x \to 0^+} \dfrac{1}{x^2} = \infty, \ \lim\limits_{x \to 0} \dfrac{1}{x^2} = \infty.$

A vertical asymptote exists at $x = 0$ because the function $\dfrac{1}{x^2}$ is undefined at $x = 0$ *and* the one-sided limits are both infinite.

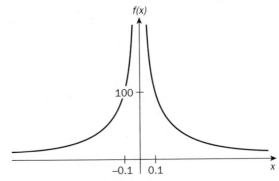

(c) Substituting $x = \sqrt{2}$ into $\dfrac{x-1}{x^2 - 2}$ results in "$\dfrac{\text{something}}{0}$." Factoring will also yield "$\dfrac{\text{something}}{0}$":

$$\lim_{x \to \sqrt{2}} \frac{x-1}{x^2 - 2} = \lim_{x \to \sqrt{2}} \frac{x-1}{\left(x+\sqrt{2}\right)\left(x-\sqrt{2}\right)} = \frac{\sqrt{2}-1}{0}$$

Conjugation will not work either:

$$\lim_{x \to \sqrt{2}} \frac{x-1}{x^2-2} = \lim_{x \to \sqrt{2}} \frac{x-1}{x^2-2} \cdot \frac{x^2+2}{x^2+2} = \lim_{x \to \sqrt{2}} \frac{(x-1)(x^2+2)}{x^4-4}$$

By computing a table of values or examining the signs of the numerator and denominator, we find:

$$\lim_{x \to \sqrt{2}^-} \frac{x-1}{x^2-2} = -\infty \quad \text{and} \quad \lim_{x \to \sqrt{2}^+} \frac{x-1}{x^2-2} = \infty$$

Because the one-sided limits are different, we say that $\lim_{x \to \sqrt{2}} \frac{x-1}{x^2-2} = \text{DNE}$, but because the one-sided limits are infinite, the function has a vertical asymptote at $x = \sqrt{2}$.

The function $f(x) = \frac{x-1}{x^2-2}$ *may* have another asymptote since the function is undefined at $x = \pm\sqrt{2}$. We've already found that there is a vertical asymptote at $x = +\sqrt{2}$. However, don't assume that the function has a vertical asymptote at $x = -\sqrt{2}$ just because the function is undefined at $x = -\sqrt{2}$. Make sure that the one-sided limits are infinite.

In this case, $\lim_{x \to -\sqrt{2}^-} \frac{x-1}{x^2-2} = -\infty$ and $\lim_{x \to -\sqrt{2}^+} \frac{x-1}{x^2-2} = \infty$, so there is also a vertical asymptote at $x = -\sqrt{2}$.

When x does not approach a fixed number but, instead, gets very, very large (i.e., $x \to \infty$) or very, very small (i.e., $x \to -\infty$), we say that the limit is approaching infinity. We use the notation $\lim_{x \to \infty} f(x)$ and $\lim_{x \to -\infty} f(x)$.

Limits Approaching Infinity: Evaluate the following limits approaching infinity:

(a) $\lim_{x \to \infty} \frac{1}{x}$ (b) $\lim_{x \to \infty} \frac{x+1}{x-3}$ (c) $\lim_{x \to -\infty} \frac{x^3+2x-9}{x^2+4x+2}$

Step 1: Factor out the highest power of x in the polynomial(s).

Step 2: Use the rules for operations with limits to simplify the function.

Step 3: Reduce to lowest terms.

Solution and Explanation:

(a) We cannot factor out the highest power of x in $\frac{1}{x}$, so we compute $\lim\limits_{x\to\infty}\frac{1}{x}$ by creating a table of values.

x	10	100	1000	1,000,000
$\frac{1}{x}$	$\frac{1}{10}=0.1$	$\frac{1}{100}=0.01$	$\frac{1}{1000}=0.001$	$\frac{1}{1,000,000}=0.000001$

As x gets bigger, $\frac{1}{x}$ approaches zero, so $\lim\limits_{x\to\infty}\frac{1}{x}=0$.

(b) We can compute $\lim\limits_{x\to\infty}\frac{x+1}{x-3}$ by factoring x out of the numerator and the denominator.

$$\lim_{x\to\infty}\frac{x+1}{x-3}=\lim_{x\to\infty}\frac{x\left(1+\frac{1}{x}\right)}{x\left(1-\frac{3}{x}\right)}$$

We've established that $\lim\limits_{x\to\infty}\frac{1}{x}=0$, so

$$\lim_{x\to\infty}\frac{x\left(1+\overset{0}{\cancel{\frac{1}{x}}}\right)}{x\left(1-\underset{0}{\cancel{\frac{3}{x}}}\right)}=\lim_{x\to\infty}\frac{\cancel{x}\left(1+0\right)}{\cancel{x}\left(1-0\right)}=1$$

We can verify this by creating a table of values.

x	10	100	1000	1,000,000
$\frac{x+1}{x-3}$	$\frac{11}{7}=1.571$	$\frac{101}{97}=1.041$	$\frac{1001}{997}=1.004$	$\frac{1,000,001}{999,997}=1.000004$

(c) To compute $\lim\limits_{x\to-\infty}\frac{x^3+2x-9}{x^2+4x+2}$, we factor out the highest power of x in the numerator and denominator.

$$\lim_{x\to-\infty}\frac{x^3+2x-9}{x^2+4x+2}=\lim_{x\to-\infty}\frac{x^3\left(1+\frac{2}{x^2}-\frac{9}{x^3}\right)}{x^2\left(1+\frac{4}{x}+\frac{2}{x^2}\right)}$$

The limits of $\frac{2}{x^2}$ and $\frac{9}{x^3}$ both equal zero because the fractions become smaller and smaller as the absolute value of the denominator gets bigger and bigger.

$$\lim_{x \to -\infty} \frac{x^3 \left(1 + \dfrac{2^{\,0}}{x^2} - \dfrac{9^{\,0}}{x^3}\right)}{x^2 \left(1 + \dfrac{4}{x}_0 + \dfrac{2}{x^2}_0\right)} = \lim_{x \to -\infty} \frac{x^3 (1 + 0 - 0)}{x^2 (1 + 0 + 0)} = \lim_{x \to -\infty} x = -\infty$$

We can also think about this limit more informally. For the function $\frac{x^3 + 2x - 9}{x^2 + 4x + 2}$, as x gets very, very large and negative, the x^3 term is controlling the numerator. The other terms in the numerator, $2x$ and 9, are insignificant. Similarly, the denominator is controlled by x^2, the term of highest degree.

For very large, negative values of x, the numerator of the function "looks like" x^3 and the denominator "looks like" x^2.

As $x \to -\infty$, the function $\frac{x^3 + 2x - 9}{x^2 + 4x + 2}$ behaves like the function $\frac{x^3}{x^2} = x$.

Therefore, $\lim_{x \to -\infty} \frac{x^3 + 2x - 9}{x^2 + 4x + 2} = \lim_{x \to -\infty} \frac{x^3}{x^2} = \lim_{x \to -\infty} x = -\infty$ as we computed previously.

Now let's apply this knowledge about limits to trigonometric functions.

Limits of Trigonometric Functions: Evaluate the limits of these trigonometric functions:

(a) $\lim_{x \to 0} \dfrac{\cos x}{\sin x - 3}$ (b) $\lim_{x \to \pi^+} \cot x$ (c) $\lim_{x \to 0} \dfrac{\sin \frac{x}{2}}{x}$ (d) $\lim_{x \to 0} \dfrac{1 - \sec x}{x}$

Step 1: Substitute the value of c into the function.

Step 2: If substitution does not work, manipulate the function into the form $\frac{\sin x}{x}$ or $\frac{\cos x - 1}{x}$.

Step 3: Simplify using the special trig limits: $\lim_{x \to 0} \frac{\sin x}{x} = 1$ $\lim_{x \to 0} \frac{\cos x - 1}{x} = 0$.

Solution and Explanation: Factoring and conjugating generally won't be helpful in computing limits involving trig functions. The sine limit, $\lim_{x \to 0} \frac{\sin x}{x} = 1$ is usually what we use when we need to compute limits involving trigonometric functions; we try to convert the trigonometric functions into $\frac{\sin \text{"something"}}{\text{"something"}}$. Another useful rule is $\lim_{x \to 0} \frac{\cos x - 1}{x} = 0$.

Note that these limits are only true when we are expressing x in radians, not degrees.

(a) To evaluate $\lim\limits_{x \to 0} \dfrac{\cos x}{\sin x - 3}$, we substitute 0 for x. Recall that $\cos 0 = 1$ and $\sin 0 = 0$.

Therefore, $\lim\limits_{x \to 0} \dfrac{\cos x}{\sin x - 3} = \lim\limits_{x \to 0} \dfrac{\cos 0}{\sin 0 - 3} = \dfrac{1}{-3}$.

(b) To evaluate $\lim\limits_{x \to \pi^+} \cot x$, we first simplify the function using the identity $\cot x = \dfrac{\cos x}{\sin x}$.

$$\lim\limits_{x \to \pi^+} \cot x = \lim\limits_{x \to \pi^+} \dfrac{\cos x}{\sin x} = \lim\limits_{x \to \pi^+} \dfrac{\cos \pi}{\sin \pi} = \dfrac{1}{0} .$$

Because our answer is " $\dfrac{\text{something}}{0}$," we are dealing with infinity. As $x \to \pi^+$, the numerator is fixed and negative. The table below shows us that the denominator is also negative.

Angle (radians)	$\dfrac{7\pi}{6}$	$\dfrac{5\pi}{4}$	$\dfrac{4\pi}{3}$	$\dfrac{3\pi}{2}$
sine	$-\dfrac{1}{2}$	$-\dfrac{\sqrt{2}}{2}$	$-\dfrac{\sqrt{3}}{2}$	-1

Therefore, $\lim\limits_{x \to \pi^+} \cot x = \infty.$

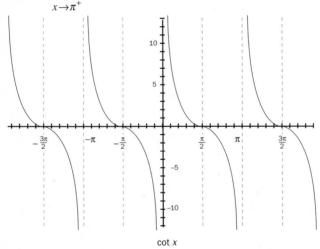

cot x

(c) We can use the rule $\lim\limits_{x\to 0}\dfrac{\sin x}{x}=1$ to evaluate $\lim\limits_{x\to 0}\dfrac{\sin\frac{x}{2}}{x}$. The key is to convert the function so that it contains $\dfrac{\sin\text{"something"}}{\text{"something"}}$. We accomplish this by multiplying both the numerator and the denominator by $\frac{1}{2}$ so that the denominator is $\frac{x}{2}$:

$$\lim_{x\to 0}\frac{\sin\frac{x}{2}}{x}=\lim_{x\to 0}\frac{\frac{1}{2}}{\frac{1}{2}}\cdot\frac{\sin\frac{x}{2}}{x}=\lim_{x\to 0}\frac{1}{2}\cdot\frac{\sin\frac{x}{2}}{\frac{x}{2}}.$$

One of the rules for operations with limits is that
$$\lim_{x\to c}\left[f(x)\cdot g(x)\right]=\left(\lim_{x\to c}f(x)\right)\left(\lim_{x\to c}g(x)\right).$$

Therefore, $\lim\limits_{x\to 0}\dfrac{1}{2}\cdot\dfrac{\sin\frac{x}{2}}{\frac{x}{2}}=\left(\lim\limits_{x\to 0}\dfrac{1}{2}\right)\left(\lim\limits_{x\to 0}\dfrac{\sin\frac{x}{2}}{\frac{x}{2}}\right)=\dfrac{1}{2}\cdot 1=\dfrac{1}{2}.$

(d) To evaluate $\lim\limits_{x\to 0}\dfrac{1-\sec x}{x}$, use the identity $\sec x=\dfrac{1}{\cos x}$.

$$\lim_{x\to 0}\frac{1-\sec x}{x}=\lim_{x\to 0}\frac{1-\dfrac{1}{\cos x}}{x}$$

$$=\lim_{x\to 0}\frac{\dfrac{\cos x-1}{\cos x}}{x}$$

$$=\lim_{x\to 0}\frac{\cos x-1}{x\cos x}$$

At this point, we can use the rule $\lim\limits_{x\to 0}\dfrac{\cos x-1}{x}=0$ by breaking up the function $\dfrac{\cos x-1}{x\cos x}$ into a product of $\left(\dfrac{1}{\cos x}\right)\left(\dfrac{\cos x-1}{x}\right)$.

$$\lim_{x\to 0}\frac{\cos x-1}{x\cos x}=\lim_{x\to 0}\left(\frac{1}{\cos x}\right)\left(\frac{\cos x-1}{x}\right)=(1)(0)=0.$$

CHAPTER QUIZ

1. Evaluate the limit and continuity of the graph at $x = b$:

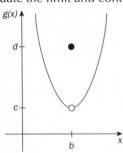

(A) $\displaystyle\lim_{x \to b} g(x) = c$ and the function is continuous at $x = b$

(B) $\displaystyle\lim_{x \to b} g(x) = c$ and the function has a point discontinuity at $x = b$

(C) $\displaystyle\lim_{x \to b} g(x) = c$ and the function has a jump discontinuity at $x = b$

(D) $\displaystyle\lim_{x \to b} g(x) = $ DNE and has a point discontinuity at $x = b$

(E) $\displaystyle\lim_{x \to b} g(x) = $ DNE and has a jump discontinuity at $x = b$

2. Suppose that *f* is *not* continuous at *x* = 1, *f* is defined at *x* = 1, and $\lim_{x \to 1} f(x) = L$, where *L* is finite. Which of the following could be the graph of *f*?

(A)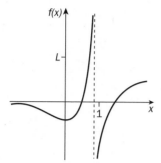

(B)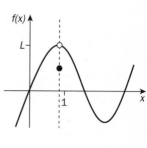

(C)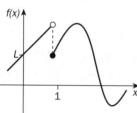

(D)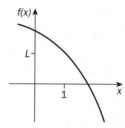

(E)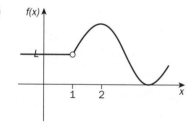

3. Evaluate $\lim_{x \to -1} \dfrac{2x^2 - x - 3}{x^2 - 2x - 3}$

(A) $\dfrac{1}{4}$

(B) 0

(C) −1

(D) $-\dfrac{1}{4}$

(E) DNE

4. Evaluate $\lim_{x \to 0} \dfrac{\dfrac{1}{x+4} - \dfrac{1}{4}}{x}$

(A) 0

(B) $\dfrac{1}{16}$

(C) $\dfrac{-1}{16}$

(D) 4

(E) DNE

5. Evaluate $\lim\limits_{x \to 2} \dfrac{x + 3}{x - 2}$

   (A) 0

   (B) 2

   (C) ∞

   (D) $-\infty$

   (E) DNE

6. Which of the following functions has a vertical asymptote at $x = 4$?

   (A) $\dfrac{x + 5}{x^2 - 4}$

   (B) $\dfrac{x^2 - 16}{x - 4}$

   (C) $\dfrac{4x}{x + 1}$

   (D) $\dfrac{x + 6}{x^2 - 7x + 12}$

   (E) None of the above.

7. Evaluate $\lim\limits_{x \to +\infty} \dfrac{2x^2 + 3}{x^2 - 5x - 1}$

   (A) 0

   (B) 1

   (C) 2

   (D) $+\infty$

   (E) DNE

8. Evaluate $\lim\limits_{x \to +\infty} \dfrac{x^3 - 2}{5x^4 - 3x^3 + 2x}$

   (A) 0

   (B) 1

   (C) 2

   (D) $+\infty$

   (E) DNE

9. Evaluate $\lim\limits_{x \to 0} \dfrac{\tan 2x}{x}$

   (A) ∞

   (B) 0

   (C) 1

   (D) 2

   (E) DNE

10. Evaluate $\lim\limits_{x \to \infty} \dfrac{\sin \frac{1}{x}}{\frac{1}{x}}$

   (A) 0

   (B) 1

   (C) -1

   (D) ∞

   (E) DNE

Answers and Explanations

1. B

At $x = b$, the left and right sides of the parabola meet at $y = c$. In limit notation, we write $\lim\limits_{x \to b^-} g(x) = c$ and $\lim\limits_{x \to b^+} g(x) = c$. Since the one-sided limits are equal, $\lim\limits_{x \to b} g(x) = c$. However, there is a hole at $x = b$, so the function has a point discontinuity at $x = b$.

The solid circle on the graph indicates that $g(b) = d$.

2. B

The function shown in option (A) has an asymptote at $x = 1$, so the function shown in this graph is not defined at $x = 1$ and it cannot be the graph of $f(x)$. The left- and right-hand limits of option (C) are not equal, so it cannot be the graph of $f(x)$. The function shown in option (D) is continuous at $x = 1$, so this cannot be the graph of $f(x)$. Option (E) is not defined at $x = 1$.

The remaining answer choice is (B), and this graph meets the conditions given for $f(x)$.

3. A

When we substitute -1 for x, we get

$$\lim_{x \to -1} \frac{2x^2 - x - 3}{x^2 - 2x - 3} = \frac{2(-1)^2 - (-1) - 3}{(-1)^2 - 2(-1) - 3} = \frac{2 + 1 - 3}{1 + 2 - 3} = \frac{0}{0}.$$

Since substitution does not work, we can try factoring instead:

$$\lim_{x \to -1} \frac{2x^2 - x - 3}{x^2 - 2x - 3} = \frac{(2x - 3)(x+1)}{(x - 3)(x+1)} = \frac{2(-1) - 3}{-1 - 3} = \frac{-1}{-4} = \frac{1}{4}$$

4. C

To evaluate $\lim\limits_{x \to 0} \dfrac{\frac{1}{x+4} - \frac{1}{4}}{x}$, we simplify the function first by adding the fractions in the numerator.

$$\lim_{x \to 0} \frac{\frac{1}{x+4} - \frac{1}{4}}{x} = \lim_{x \to 0} \frac{\frac{4 - (x+4)}{4(x+4)}}{x} = \lim_{x \to 0} \frac{4 - x - 4}{4x(x+4)} = \lim_{x \to 0} \frac{-x}{4x(x+4)} = \lim_{x \to 0} \frac{-1}{4x + 16} = \frac{-1}{16}.$$

5. E

Substitution gives us $\lim\limits_{x\to 2}\dfrac{x+3}{x-2}=\dfrac{5}{0}$, so we know that we are dealing with infinity.

$$\lim_{x\to 2^-}\frac{x+3}{x-2}=-\infty:$$

x	1.9	1.99	1.999
$\dfrac{x+3}{x-2}$	$\dfrac{1.9+3}{1.9-2}=-49$	$\dfrac{1.99+3}{1.99-2}=-499$	$\dfrac{1.999+3}{1.999-2}=-4999$

$$\lim_{x\to 2^+}\frac{x+3}{x-2}=\infty$$

x	2.1	2.01	2.001
$\dfrac{x+3}{x-2}$	$\dfrac{2.1+3}{2.1-2}=51$	$\dfrac{2.01+3}{2.01-2}=501$	$\dfrac{2.001+3}{2.001-2}=5001$

$\lim\limits_{x\to 2}\dfrac{x+3}{x-2}=$ DNE because the one-sided limits are not equal. However, a vertical asymptote does exist at $x=2$ because the one-sided limits are both infinity.

6. D

We can eliminate choice (A) immediately because the only candidates for the vertical asymptotes of this function are the points where the denominator equals zero—at $x=\pm 2$.

We can also eliminate choice (C) because the only candidate for a vertical asymptote of this function is $x=-1$.

In choice (B), $x=4$ is a candidate for a vertical asymptote because the function is not defined at this point. To determine whether this point is in fact a vertical asymptote, we compute the limit:

$$\lim_{x\to 4}\frac{x^2-16}{x-4}=\lim_{x\to 4}\frac{(x+4)\,\cancel{(x-4)}}{\cancel{x-4}}=8.$$

Because the limit is finite, the function does not have a vertical asymptote at $x=4$.

We can rewrite the function in (D) and see that $x = 4$ is a candidate for a vertical asymptote because the function is not defined at this point:

$$\frac{x+6}{x^2 - 7x + 12} = \frac{x+6}{(x-3)(x-4)}$$

We compute the limit:

$$\lim_{x \to 4^+} \frac{x+6}{(x-3)(x-4)} = \frac{10}{1 \cdot \text{something small and positive}} = +\infty$$

$$\lim_{x \to 4^-} \frac{x+6}{(x-3)(x-4)} = \frac{10}{1 \cdot \text{something small and negative}} = -\infty$$

Because the one-sided limits are infinite, this function has a vertical asymptote at $x = 4$ and the correct answer is (D).

7. C

We factor out the largest power of x in the numerator and the denominator.

$$\lim_{x \to +\infty} \frac{2x^2 + 3}{x^2 - 5x - 1} = \lim_{x \to +\infty} \frac{x^2 \left(2 + \cancelto{0}{\frac{3}{x^2}} \right)}{x^2 \left(1 - \cancelto{0}{\frac{5}{x}} - \cancelto{0}{\frac{1}{x^2}} \right)} = \frac{2}{1} = 2$$

8. A

We can factor out the largest power of x in the numerator and denominator, but we can also compute the limit by reducing the polynomials to their terms of highest degree.

The numerator "looks like" x^3 and the denominator "looks like" $5x^4$.

$$\lim_{x \to +\infty} \frac{x^3 - 2}{5x^4 - 3x^3 + 2x} = \lim_{x \to +\infty} \frac{x^3}{5x^4} = \lim_{x \to +\infty} \frac{1}{5x} = 0.$$

9. D

To evaluate $\lim\limits_{x \to 0} \frac{\tan 2x}{x}$, we can try plugging in first, but that yields the indeterminate form $\frac{0}{0}$. Next, we use the identity $\tan x = \frac{\sin x}{\cos x}$ to set ourselves up to use the formula $\lim\limits_{x \to 0} \frac{\sin x}{x} = 1$.

$$\lim_{x \to 0} \frac{\tan 2x}{x} = \lim_{x \to 0} \frac{\frac{\sin 2x}{\cos 2x}}{x} = \lim_{x \to 0} \frac{\sin 2x}{x \cos 2x}$$

We multiply the numerator and denominator by 2 to get a $2x$ in the denominator.

$$\lim_{x \to 0} \frac{\sin 2x}{x \cos 2x} = \lim_{x \to 0} \frac{2}{2} \cdot \frac{\sin 2x}{x \cos 2x} = \lim_{x \to 0} \frac{\sin 2x}{2x} \cdot \frac{2}{\cos 2x} = \lim_{x \to 0} \frac{2}{\cos 2x}$$

We evaluate $\lim\limits_{x \to 0} \frac{2}{\cos 2x}$ by plugging in 0 for x.

$$\lim_{x \to 0} \frac{2}{\cos 2x} = 2$$

10. B

This limit is a variation on the limit $\lim\limits_{x \to 0} \frac{\sin x}{x} = 1$. Notice that if $x \to \infty$, then $\frac{1}{x} \to 0$; therefore, this limit is just a fancier way of writing $\lim\limits_{x \to 0} \frac{\sin x}{x}$.

Derivatives

WHAT ARE DERIVATIVES?

Derivatives are one of the two main concepts in calculus (the other one is integrals, which we will discuss in Chapters 9 and 10). Derivatives can be understood in two basic ways: physically, they are a rate of change; graphically, they are the slope of the line tangent to the curve at a point. The notation for the derivative of a function $f(x)$ can be written as $\frac{d}{dx}(f(x)), \frac{df}{dx},$ or $f'(x)$.

CONCEPTS TO HELP YOU

1. Difference quotient: The derivative is defined as the limit of the difference quotient:

$$\frac{d}{dx}f(x) = \lim_{\Delta x \to 0} \frac{f(x_0 + \Delta x) - f(x_0)}{\Delta x} \quad \text{or} \quad \frac{d}{dx}f(x) = \lim_{h \to 0} \frac{f(x_0 + h) - f(x_0)}{h}$$

2. Table of derivative formulas: Derivative formulas are shortcuts to using the limit of the difference quotient. It is a good idea to memorize the following formulas:

Name	Formula
Derivative of a Constant	$\frac{d}{dx}(f(x)) = \frac{d}{dx}(c) = 0$
Power Rule	$\frac{d}{dx}(x^n) = nx^{n-1}$, n an integer
Derivative of $\sqrt{x}$	$\frac{d}{dx}(\sqrt{x}) = \frac{1}{2\sqrt{x}}$
Constant Multiple Rule	$\frac{d}{dx}[cf(x)] = c \cdot \frac{d}{dx}(f(x))$, c a real number

Sum and Difference Rule	$\dfrac{d}{dx}(f(x)+g(x)) = \dfrac{d}{dx}(f(x)) + \dfrac{d}{dx}(g(x))$ for f and g differentiable at x
Product Rule	$\dfrac{d}{dx}(f(x) \cdot g(x)) = \dfrac{d}{dx}(f(x)) \cdot g(x) + f(x) \cdot \dfrac{d}{dx}(g(x))$ for f and g differentiable at x
Quotient Rule	$\dfrac{d}{dx}\left(\dfrac{f(x)}{g(x)}\right) = \dfrac{\dfrac{d}{dx}(f(x)) \cdot g(x) - f(x) \cdot \dfrac{d}{dx}(g(x))}{(g(x))^2}$ for f and g differentiable at x, $g(x) \neq 0$
Chain Rule	$\dfrac{d}{dx}h(g(x)) = \dfrac{dh}{dx}(g(x)) \cdot \dfrac{d}{dx}g(x)$ for g differentiable at x and h differentiable at $g(x)$
Derivatives of Trigonometric Functions	$\dfrac{d}{dx}(\sin x) = \cos x$ $\qquad$ $\dfrac{d}{dx}(\cos x) = -\sin x$ $\dfrac{d}{dx}(\tan x) = \sec^2 x$ $\qquad$ $\dfrac{d}{dx}(\cot) = -\csc^2 x$ $\dfrac{d}{dx}(\sec x) = \sec x \tan x$ $\qquad$ $\dfrac{d}{dx}(\sec x) = -\csc x \cot x$

STEPS YOU NEED TO REMEMBER

1. Break up the derivative.

To compute the derivative of any polynomial, use the Sum and Difference Rule to break the derivative into a sum and difference of derivatives. For example,

$$\frac{d}{dx}\left(8x^4 - 1.2x^2 + \frac{2}{3}x + \pi\right) = \frac{d}{dx}\left(8x^4\right) - \frac{d}{dx}\left(1.2x^2\right) + \frac{d}{dx}\left(\frac{2}{3}x\right) + \frac{d}{dx}(\pi)$$

2. Pull the constant out of each derivative.

Use the Constant Multiple Rule to pull the constant out of each derivative:

$$\frac{d}{dx}\left(8x^4\right) - \frac{d}{dx}\left(1.2x^2\right) + \frac{d}{dx}\left(\frac{2}{3}x\right) + \frac{d}{dx}(\pi) = 8\frac{d}{dx}\left(x^4\right) - 1.2\frac{d}{dx}\left(x^2\right) + \frac{2}{3}\frac{d}{dx}(x) + \frac{d}{dx}(\pi)$$

3. Differentiate each term.

Use the Power Rule and Constant Rule to differentiate each term:

$$8\frac{d}{dx}\left(x^4\right)-1.2\frac{d}{dx}\left(x^2\right)+\frac{2}{3}\frac{d}{dx}(x)+\frac{d}{dx}(\pi)=8\cdot4x^3-1.2\cdot2x+\frac{2}{3}+0$$

4. Simplify.

Simplify by combining like terms:

$$8\cdot4x^3-1.2\cdot2x+\frac{2}{3}+0=32x^3-2.4x+\frac{2}{3}$$

COMMON DERIVATIVE QUESTIONS

Limit of the Difference Quotient: Compute the derivative as the limit of the difference quotient:

(a) $f(x)=4x-5$ (b) $f(x)=x^2-3x+4$

Step 1: Replace the x in the function with the expression (x_0+h).

Step 2: Substitute the function for the expression $f(x_0+h)$ in the difference quotient.

Step 3: Replace the x in the function with the expression x_0.

Step 4: Substitute the function for $f(x_0)$ in the difference quotient.

Step 5: Simplify the difference quotient.

Step 6: Evaluate the limit.

Solution and Explanation: We can interpret the definition of the derivative as an instantaneous rate of change geometrically, on a graph.

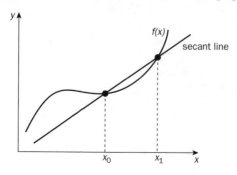

The rate of change of y between two points x_0 and x_1 is the slope of the secant line and is given by the difference quotient $\frac{f(x_1)-f(x_0)}{x_1-x_0}$, which is simply another way of writing the slope formula $\frac{y_1-y_0}{x_1-x_0}$.

The distance between x_0 and x_1 is the change in x and is written as Δx ("delta x"). The point x_1, therefore, can be written as $x_0 + \Delta x$.

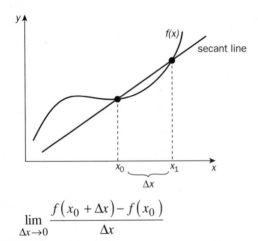

$$\lim_{\Delta x \to 0} \frac{f(x_0 + \Delta x) - f(x_0)}{\Delta x}$$

As the points x_0 and x_1 get closer and closer together, it is hard to distinguish between the two points x_0 and x_1, or, using the notation above, x_0 and $x_0 + \Delta x$. As $\Delta x \to 0$, the points run together and the secant line between the points becomes tangent to the curve, i.e., it is the line that best approximates the direction of the curve. We call this the tangent line to the graph at the point x_0.

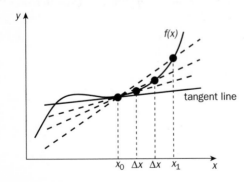

The slope of the tangent line to the graph at x_0 is the limit of the slopes of the secant lines between x_0 and $x_0 + \Delta x$ for Δx close to zero. This slope is given by the limit of the difference quotient: $\displaystyle\lim_{\Delta x \to 0} \frac{f(x_0 + \Delta x) - f(x_0)}{\Delta x}$.

Sometimes, the expression Δx is replaced by h, where h, like Δx, is a number close to zero. In this case, the difference quotient becomes $\dfrac{f(x_0 + h) - f(x_0)}{h}$.

(a) To compute the derivative of $f(x) = 4x - 5$ as the limit of the difference quotient, we replace the x in the function with $(x_0 + h)$ and x_0:

$$\lim_{h \to 0} \frac{f(x_0 + h) - f(x_0)}{h} = \lim_{h \to 0} \frac{\left[4(x_0 + h) - 5\right] - \left[4x_0 - 5\right]}{h}$$

$$= \lim_{h \to 0} \frac{4x_0 + 4h - 5 - 4x_0 + 5}{h}$$

$$= \lim_{h \to 0} \frac{4h}{h}$$

$$= \lim_{h \to 0} 4$$

$$= 4$$

(b) The derivative of $f(x) = x^2 - 3x + 4$ as the limit of the difference quotient is

$$\lim_{h \to 0} \frac{f(x_0 + h) - f(x_0)}{h} = \lim_{h \to 0} \frac{\left[(x_0 + h)^2 - 3(x_0 + h) + 4\right] - \left[(x_0)^2 - 3(x_0) + 4\right]}{h}$$

$$= \lim_{h \to 0} \frac{\cancel{x_0^2} + 2x_0 h + h^2 - \cancel{3x_0} - 3h + \cancel{4} - \cancel{x_0^2} + \cancel{3x_0} - \cancel{4}}{h}$$

$$= \lim_{h \to 0} \frac{2x_0 h + h^2 - 3h}{h}$$

$$= \lim_{h \to 0} \frac{h(2x_0 + h - 3)}{h}$$

$$= \lim_{h \to 0} (2x_0 + h - 3)$$

$$= 2x_0 - 3$$

As you can see, computing the derivative as the limit of the difference quotient can get quite complicated. Fortunately, we have the derivative formulas.

Derivatives of Polynomials: Compute the derivatives using the derivative formulas:

(a) $\frac{d}{dx}\left(4x^3 - 5x + 4\right)$ (b) $\frac{d}{dx}\left[\left(x^3 + 7\right)\left(5x^2 - \frac{3}{2}\right)\right]$ (c) $\frac{d}{dx}\left(\frac{3x+5}{2x-3}\right)$

Step 1: Determine which derivative formula to use:

- If the function is a single polynomial, use the Sum and Difference Rule.
- If the function is a product of two polynomials, plug the polynomials into the Product Rule.
- If the function is a quotient of two polynomials, plug the polynomials into the Quotient Rule.

Step 2: Follow the procedure for computing the derivative of any polynomial:

1) Break up the derivative.
2) Pull the constant out of each derivative.
3) Differentiate each term.
4) Simplify.

Solution and Explanation:

(a) We can compute the derivative of $(4x^3 - 5x + 4)$ by using the Constant Multiple Rule and the Sum and Difference Rule.

The Constant Multiple Rule is $\frac{d}{dx}\left[c \cdot f(x)\right] = c \cdot \frac{d}{dx} f(x)$.

The Sum and Difference Rule is $\frac{d}{dx}\left(f(x) + g(x)\right) = \frac{d}{dx}\left(f(x)\right) + \frac{d}{dx}\left(g(x)\right)$.

First, we use the Sum and Difference Rule to break the derivative into a sum and difference of derivatives:

$$\frac{d}{dx}\left(4x^3 - 5x + 4\right) = \frac{d}{dx}\left(4x^3\right) - \frac{d}{dx}(5x) + \frac{d}{dx}(4)$$

Next, we use the Constant Multiple Rule to pull the constant out of each derivative:

$$\frac{d}{dx}\left(4x^3\right) - \frac{d}{dx}(5x) + \frac{d}{dx}(4) = 4\frac{d}{dx}\left(x^3\right) - 5\frac{d}{dx}(x) + \frac{d}{dx}(4)$$

Then, we use the Power Rule and the Constant Rule to differentiate each term:

$$4\frac{d}{dx}\left(x^3\right)-5\frac{d}{dx}\left(x\right)+\frac{d}{dx}\left(4\right)=4\left(3x^{3-1}\right)-5\left(x^{1-1}\right)+0$$

$$=12x^2-5+0$$

$$=12x^2-5$$

Therefore, $\dfrac{d}{dx}\left(4x^3-5x+4\right)=12x^2-5$

(b) To find the derivative of the function $\left(x^3+7\right)\left(5x^2-\dfrac{3}{2}\right)$, we use the Product Rule.

The Product Rule is

$$\frac{d}{dx}\left(f(x)\cdot g(x)\right)=\frac{d}{dx}\left(f(x)\right)\cdot g(x)+f(x)\cdot\frac{d}{dx}\left(g(x)\right).$$

Let $f(x)=(x^3+7)$ and $g(x)=\left(5x^2-\dfrac{3}{2}\right)$.

$$\frac{d}{dx}\left(x^3+7\right)=\frac{d}{dx}\left(x^3\right)+\frac{d}{dx}\left(7\right) \qquad \frac{d}{dx}\left(5x^2-\frac{3}{2}\right)=\frac{d}{dx}\left(5x^2\right)-\frac{d}{dx}\left(\frac{3}{2}\right)$$

$$=\left(3x^{3-1}\right)+0 \qquad\text{and}\qquad =5\frac{d}{dx}\left(x^2\right)-\frac{d}{dx}\left(\frac{3}{2}\right)$$

$$=3x^2 \qquad\qquad =5\left(2x^{2-1}\right)-0$$

$$=10x$$

Therefore,

$$\frac{d}{dx}\left[\left(x^3+7\right)\left(5x^2-\frac{3}{2}\right)\right]=\frac{d}{dx}\left(x^3+7\right)\cdot\left(5x^2-\frac{3}{2}\right)+\left(x^3+7\right)\cdot\frac{d}{dx}\left(5x^2-\frac{3}{2}\right)$$

$$=\left(3x^2\right)\left(5x^2-\frac{3}{2}\right)+\left(x^3+7\right)\left(10x\right)$$

$$=\left(15x^4-\frac{9}{2}x^2\right)+\left(10x^4+70x\right)$$

$$=25x^4+70x-\frac{9}{2}x^2$$

We can also compute the derivative of f by multiplying and then computing the derivative of the resulting polynomial:

$$\frac{d}{dx}\left[\left(x^3+7\right)\left(5x^2-\frac{3}{2}\right)\right]=\frac{d}{dx}\left(5x^5-\frac{3}{2}x^3+35x^2-\frac{21}{2}\right)$$

$$=25x^4-\frac{9}{2}x^2+70x$$

Both methods give the same answer, which is what we expect. In general, if you can multiply something out or do some other algebraic manipulation to make the problem simpler *before* computing the derivative, it's a good idea to go ahead and do it.

(c) To compute the derivative of $\left(\dfrac{3x+5}{2x-3}\right)$, we use the Quotient Rule.

The Quotient Rule is

$$\frac{d}{dx}\left(\frac{f(x)}{g(x)}\right) = \frac{\frac{d}{dx}(f(x))\cdot g(x)-f(x)\cdot\frac{d}{dx}(g(x))}{(g(x))^2}.$$

Let $f(x) = (3x+5)$ and $g(x) = (2x-3)$

$$\frac{d}{dx}f(x)=\frac{d}{dx}(3x+5) \qquad\qquad \frac{d}{dx}g(x)=\frac{d}{dx}(2x-3)$$
$$=\frac{d}{dx}(3x)+\frac{d}{dx}(5) \quad\text{and}\qquad =\frac{d}{dx}(2x)-\frac{d}{dx}(3)$$
$$=3+0 \qquad\qquad\qquad\qquad =2-0$$
$$=3 \qquad\qquad\qquad\qquad\qquad =2$$

Therefore,

$$\frac{d}{dx}\left(\frac{f(x)}{g(x)}\right) = \frac{\frac{d}{dx}(f(x))\cdot g(x)-f(x)\cdot\frac{d}{dx}(g(x))}{(g(x))^2}$$
$$=\frac{3(2x-3)-(3x+5)(2)}{(2x-3)^2}$$
$$=\frac{6x-9-6x-10}{(2x-3)^2}$$
$$=\frac{-19}{(2x-3)^3}$$

We can also find the derivatives of trigonometric functions by applying the same rules and formulas.

Derivatives of Trigonometirc Functions: Use the Quotient Rule to prove that:

(a) $\dfrac{d}{dx}\tan x = \sec^2 x$ (b) $\dfrac{d}{dx}\cot x = -\csc^2 x$

(c) $\dfrac{d}{dx}\sec x = \sec x \tan x$ (d) $\dfrac{d}{dx}\csc x = -\csc x \cot x$

Step 1: Use reciprocal and ratio identities to simplify the function into sine and cosine.

Step 2: Substitute the function into the Quotient Rule. At the very least, you should memorize the derivatives of sine and cosine:

- $\dfrac{d}{dx}\sin x = \cos x$

- $\dfrac{d}{dx}\cos x = -\sin x$

Step 3: Simplify and apply other trigonometric identities where appropriate.

Solution and Explanation:

(a) Recall that $\tan x = \dfrac{\sin x}{\cos x}$.

$$\frac{d}{dx}\tan x = \frac{d}{dx}\left(\frac{\sin x}{\cos x}\right)$$

$$= \frac{\dfrac{d}{dx}\sin x \cdot \cos x - \sin x \cdot \dfrac{d}{dx}\cos x}{\cos^2 x}$$

$$= \frac{\cos x \cdot \cos x - (\sin x)(-\sin x)}{\cos^2 x}$$

$$= \frac{\cos^2 x + \sin^2 x}{\cos^2 x}$$

Because $\cos^2 x + \sin^2 x = 1$ (Pythagorean identity),

$$\frac{d}{dx}\tan x = \frac{\cos^2 x + \sin^2 x}{\cos^2 x}$$

$$= \frac{1}{\cos^2 x}$$

$$= \sec^2 x$$

(b) Recall that $\cot x = \dfrac{\cos x}{\sin x}$.

$\dfrac{d}{dx}\cot x = \dfrac{d}{dx}\dfrac{\cos x}{\sin x}$

$= \dfrac{\dfrac{d}{dx}\cos x \cdot (\sin x) - (\cos x) \cdot \dfrac{d}{dx}\sin x}{\sin^2 x}$

$= \dfrac{(-\sin x)(\sin x) - \cos x(\cos x)}{\sin^2 x}$

$= \dfrac{-\sin^2 x - \cos^2 x}{\sin^2 x}$

$= \dfrac{-\left(\sin^2 x + \cos^2 x\right)}{\sin^2 x}$

$= \dfrac{-1}{\sin^2 x}$

$= -\csc^2 x$

(c) Recall that $\sec x = \dfrac{1}{\cos x}$.

$\dfrac{d}{dx}\sec x = \dfrac{d}{dx}\left(\dfrac{1}{\cos x}\right)$

$= \dfrac{\dfrac{d}{dx}(1) \cdot \cos x - (1) \cdot \dfrac{d}{dx}\cos x}{\cos^2 x}$

$= \dfrac{(0)(\cos x) - (1)(-\sin x)}{\cos^2 x}$

$= \dfrac{\sin x}{(\cos x)(\cos x)}$

$= \dfrac{1}{\cos x} \cdot \dfrac{\sin x}{\cos x}$

$= \sec x \tan x$

(d) Recall that $\csc x = \dfrac{1}{\sin x}$

$$\frac{d}{dx}\csc x = \frac{d}{dx}\left(\frac{1}{\sin x}\right)$$

$$= \frac{\dfrac{d}{dx}(1)\cdot\sin x - (1)\cdot\dfrac{d}{dx}\sin x}{\sin^2 x}$$

$$= \frac{0 - \cos x}{(\sin x)(\sin x)}$$

$$= \frac{-1}{\sin x}\cdot\frac{\cos x}{\sin x}$$

$$= -\csc x \cot x$$

For more complicated functions, such as the ones we'll see in the following problem, we use the Chain Rule. The Chain Rule is one way of simplifying the derivative of functions by treating the function as a composite of functions.

In Chapter 3, we learned that a composite function is made up of two functions $f(x)$ and $g(x)$ and is written as $(f \circ g)(x)$ or $f(g(x))$.

Derivatives of Composite Functions: Compute the derivative of the composite functions:

(a) $\dfrac{d}{dx}\left(x^2 + 4x + 7\right)^{14}$ (b) $\dfrac{d}{dx}\sin x^2$ (c) $\dfrac{d}{dx}\sqrt{\sin x^2}$

Step 1: Break up the function into an inside function and an outside function.

Step 2: Substitute the functions into the Chain Rule.

Step 3: Simplify.

Solution and Explanation: The Chain Rule, as its name implies, is a chain of steps for computing the derivative of composite functions. Using this rule, we break up the original function $f(x)$ into an inside function $g(x)$ and an outside function $h(x)$: $f(x) = h(g(x))$.

The Chain Rule is $\dfrac{d}{dx}\left(h\left(g\left(x\right)\right)\right) = \dfrac{dh}{dx}\left(g\left(x\right)\right)\cdot\dfrac{d}{dx}g\left(x\right)$.

(a) The function $f(x) = (x^2 + 4x + 7)^{14}$ can be broken up into the inside function $g(x) = x^2 + 4x + 7$ and the outside function $h(x) = x^{14}$.

The derivative of the outside function is $\dfrac{d}{dx} h(x) = \dfrac{d}{dx}\left(x^{14}\right) = 14x^{13}$.

The derivative of the outside function evaluated at the inside function is $\dfrac{dh}{dx}(g(x)) = 14\left(x^2 + 4x + 7\right)^{13}$.

The derivative of the inside function is $\dfrac{d}{dx} g(x) = \dfrac{d}{dx}\left(x^2 + 4x + 7\right) = 2x + 4$.

We plug this information into the Chain Rule:

$$\frac{d}{dx}\left(h(g(x))\right) = \frac{dh}{dx}(g(x)) \cdot \frac{d}{dx} g(x)$$

$$= 14\left(x^2 + 4x + 7\right)^{13} \cdot 2x + 4$$

(b) The function $f(x) = \sin x^2$ is made up of two functions. The outside function $h(x)$ is $\sin x$ and the inside function $g(x)$ is x^2.

The derivative of the outside function is $\dfrac{d}{dx} h(x) = \dfrac{d}{dx} \sin x = \cos x$.

The derivative of the outside function evaluated at the inside function is $\dfrac{dh}{dx} g(x) = \cos x^2$.

The derivative of the inside function is $\dfrac{d}{dx} g(x) = \dfrac{d}{dx}\left(x^2\right) = 2x$.

We plug this information into the Chain Rule:

$$\frac{d}{dx}\left(h(g(x))\right) = \frac{dh}{dx}(g(x)) \cdot \frac{d}{dx} g(x)$$

$$= \cos x^2 \cdot 2x$$

$$= 2x \cos x^2$$

(c) The function $f(x) = \sqrt{\sin x^2}$ needs to be broken up into three functions: $\sqrt{x}$, $\sin x$, and x^2.

When this is the case, we use the formula

$$\frac{d}{dx}\left(t(s(r(x)))\right) = \frac{dt}{dx}(s(r(x))) \cdot \frac{ds}{dx}(r(x)) \cdot \frac{d}{dx} r(x)$$

where $t(x)$ is the outside function, $s(x)$ is the middle function, and $r(x)$ is the inside function.

For $f(x) = \sqrt{\sin x^2}$, $t(x) = \sqrt{x}$, $s(x) = \sin x$, $r(x) = x^2$, and $s(r(x)) = \sin x^2$.

The derivative of the outside function is

$$\frac{d}{dx} t(x) = \frac{d}{dx}\left(\sqrt{x}\right) = \frac{d}{dx}\left(x^{\frac{1}{2}}\right) = \frac{1}{2} \cdot x^{-\frac{1}{2}} = \frac{1}{2\sqrt{x}}$$

The derivative of the outside function evaluated at the composite of the middle and inside functions is $\frac{dt}{dx}(s(r(x))) = \dfrac{1}{2\sqrt{\sin x^2}}$.

The derivative of the middle function is $\frac{d}{dx} s(x) = \frac{d}{dx}\sin x = \cos x$.

The derivative of the middle function evaluated at the inside function is $\frac{ds}{dx} r(x) = \cos x^2$.

The derivative of the inside function is $\frac{d}{dx} r(x) = \frac{d}{dx} x^2 = 2x$.

We plug this information into the Chain Rule:

$$\frac{d}{dx}\left(t(s(r(x)))\right) = \frac{dt}{dx}(s(r(x))) \cdot \frac{ds}{dx}(r(x)) \cdot \frac{d}{dx} r(x)$$

$$= \frac{1}{2\sqrt{\sin x^2}} \cdot \cos x^2 \cdot 2x$$

$$= \frac{2x\cos x^2}{2\sqrt{\sin x^2}}$$

$$= \frac{x\cos x^2}{\sqrt{\sin x^2}}$$

Until now, the functions we have been differentiating can be defined *explicitly*, meaning that we say what one variable or value of a function is equal to in terms of other variables. For example, $y = 2x + 3$ and $f(x) = \sin^2 x$ are explicitly defined functions.

A mathematical relationship can also be defined *implicitly*, meaning that we are given an equation relating the variables, but the equation is *not* solved for one of the variables. The equation $x^2 + y^2 = 2$ is an example of an implicitly defined relationship.

In the following problem, we will compute the derivative of implicitly defined mathematical equations.

Implicit Differentiation: Solve by implicit differentiation.

(a) $x^2 + y^2 = 2$ (b) $y \sin xy^2 = \cos x$

Step 1: Differentiate both sides of the equation.

Step 2: Use the Chain Rule to differentiate terms containing the y variable.
Use the notation $\dfrac{dy}{dx}$ to denote "the derivate of y with respect to x."

Solution and Explanation: We think of x as the independent variable and y as a function of x, so the derivative of y is $\dfrac{dy}{dx}$. We differentiate both sides of the equation that relates y and x and we find an implicit relationship between x, y, and $\dfrac{dy}{dx}$.

(a) For $x^2 + y^2 = 2$, we take the derivative of both sides of the equation with respect to x, keeping in mind that, with respect to x, the derivative of y is $\dfrac{dy}{dx}$.

$$\frac{d}{dx}\left(x^2 + y^2\right) = \frac{d}{dx}(2)$$

$$\frac{d}{dx}\left(x^2\right) + \frac{d}{dx}\left(y^2\right) = 0$$

We treat (y^2) as a composite function—the inside function $g(x)$ is y and the outside function $h(x)$ is y^2.

The derivative of the outside function is $\dfrac{d}{dx}(h(x)) = \dfrac{d}{dx}\left(y^2\right) = 2y$.

The derivative of the outside function evaluated at the inside function is $\dfrac{dh}{dx}(g(x)) = 2y$.

The derivative of the inside function is $\dfrac{d}{dx}(y) = \dfrac{dy}{dx}$.

Using the Chain Rule, $\dfrac{d}{dx}\left(y^2\right) = 2y \cdot \dfrac{dy}{dx}$.

Therefore,

$$\frac{d}{dx}\left(x^2 + y^2\right) = \frac{d}{dx}(2)$$

$$\frac{d}{dx}\left(x^2\right) + \frac{d}{dx}\left(y^2\right) = 0$$

$$2x + 2y \cdot \frac{dy}{dx} = 0$$

$$2y \cdot \frac{dy}{dx} = -2x$$

$$\frac{dy}{dx} = \frac{-2x}{2y} = \frac{-x}{y}$$

We could also have started by solving the equation explicitly for y, getting $y = \sqrt{2 - x^2}$, and computing the derivative directly. Regardless of the approach we choose, the answer would be the same.

(b) For the equation $y \sin xy^2 = \cos x$, we can't find an explicit relationship between the variables. In this case, we must use implicit differentiation to find $\frac{dy}{dx}$.

We take the derivative of both sides of the equation with respect to x:

$$\frac{d}{dx}\left(y \sin xy^2\right) = \frac{d}{dx}(\cos x)$$

The derivative of the left hand side with respect to x is

$$\frac{d}{dx}\left(y \sin xy^2\right) = y \frac{d}{dx}\left(\sin xy^2\right)$$

$$= y \cdot \cos xy^2 \frac{d}{dx}\left(xy^2\right)$$

$$= y \cdot \cos xy^2 \left(y^2 + 2y \cdot \frac{dy}{dx}\right)$$

The derivative of the right hand side with respect to x is $\frac{d}{dx}(\cos x) = -\sin x$.

Therefore,

$$\frac{d}{dx}\left(y\sin xy^2\right) = \frac{d}{dx}\left(\cos x\right)$$

$$y \cdot \cos xy^2\left(y^2 + 2y \cdot \frac{dy}{dx}\right) = -\sin x$$

$$y^2 + 2y \cdot \frac{dy}{dx} = \frac{-\sin x}{y \cdot \cos xy^2}$$

$$2y \cdot \frac{dy}{dx} = \frac{-\sin x}{y \cdot \cos xy^2} - y^2$$

$$\frac{1}{2y}\left(2y \cdot \frac{dy}{dx}\right) = \frac{1}{2y}\left(\frac{-\sin x}{y \cdot \cos xy^2} - y^2\right)$$

$$\frac{dy}{dx} = \frac{-\sin x}{2y^2 \cdot \cos xy^2} - \frac{y}{2}$$

CHAPTER QUIZ

1. Which of the following expressions gives the derivative of $f(x) = \cos 2x$?

   (A) $\lim\limits_{x \to 0} \dfrac{\cos 2x}{2x}$

   (B) $\lim\limits_{x \to 0} \dfrac{\cos 2(x + h) - \cos 2x}{2x}$

   (C) $\lim\limits_{h \to 0} \dfrac{\cos 2(x + h) - \cos 2x}{h}$

   (D) $\lim\limits_{h \to 0} \dfrac{\cos(2x + h) - \cos(2x - h)}{h}$

   (E) $\lim\limits_{h \to 0} \dfrac{\cos 2x - \cos h}{h}$

2. Evaluate $\dfrac{d}{dx}\left(x^5 - \sqrt{x} + \dfrac{1}{x^3}\right)$.

   (A) $5x^4 - 2\sqrt{x} - \dfrac{3}{x^4}$

   (B) $5x^4 - \dfrac{1}{2\sqrt{x}} - \dfrac{3}{x^4}$

   (C) $5x^4 - \dfrac{1}{2\sqrt{x}} - 3x^4$

   (D) $5x^4 - \dfrac{1}{\sqrt{x}} - \dfrac{3}{x^4}$

   (E) $5x^4 - \dfrac{1}{2\sqrt{x}} - x^4$

3. Evaluate
   $\dfrac{d}{dx}\big((x + 7)(2x^2 - 3x + 5)\big)$.

   (A) $6x^2 + 22x - 16$

   (B) $x^2 + 22x - 16$

   (C) $6x^2 + x - 16$

   (D) $6x^2 + 22x + 16$

   (E) $6x^2 - 22x - 16$

4. Evaluate $\dfrac{d}{dx}\left(\dfrac{1}{x + 2}\right)$.

   (A) $\dfrac{1}{(x + 2)^2}$

   (B) $\dfrac{1}{(x + 2)}$

   (C) $\dfrac{-1}{(x + 2)}$

   (D) $\dfrac{-1}{(x + 2)^2}$

   (E) $(x + 2)^2$

5. Evaluate $\dfrac{d}{dx}\left(x^3 \cot x\right)$.

   (A) $3x^2 \cot x$

   (B) $x^3 \csc^2 x$

   (C) $3x^2 \cot x + x^3 \csc^2 x$

   (D) $\cot x - \csc^2 x$

   (E) $3x^2 \cot x - x^3 \csc^2 x$

6. Evaluate $\dfrac{d}{dx}\left(\dfrac{\cos x}{\sin 2x}\right)$.

   (A) $-2\csc x \cot x$

   (B) $\dfrac{1}{2}\csc x \cot x$

   (C) $-\dfrac{1}{2}\csc x \cot x$

   (D) $-\dfrac{1}{2}\sec x \tan x$

   (E) $-2\csc x \cot x$

7. Evaluate $\dfrac{d}{dx} \sqrt{5x^2 + 3x - 1}$.

(A) $\dfrac{10x + 3}{2\sqrt{5x^2 + 3x - 1}}$

(B) $\dfrac{1}{2\sqrt{5x^2 + 3x - 1}}$

(C) $\dfrac{10x + 3}{2\sqrt{10x + 3}}$

(D) $\dfrac{10x + 3}{\sqrt{5x^2 + 3x - 1}}$

(E) $\dfrac{\sqrt{5x^2 + 3x - 1}}{10x + 3}$

8. Evaluate $\dfrac{d}{dx} \tan(\sec x)$.

(A) $\sec x \tan x \sec^2 x$

(B) $\sec x \tan x \left(\sec^2 (\sec x) \right)$

(C) $\sec x \tan x$

(D) $\sec^2 (\sec x)$

(E) $\sec x \left(\sec^2 (\sec x) \right)$

9. Evaluate the derivative of $x^2 + 4y = xy$ and solve for $\dfrac{dy}{dx}$.

(A) $\dfrac{dy}{dx} = \dfrac{2x}{x - 4}$

(B) $\dfrac{dy}{dx} = \dfrac{2x - y}{x}$

(C) $\dfrac{dy}{dx} = \dfrac{x - 4}{2x - y}$

(D) $\dfrac{dy}{dx} = \dfrac{2x - y}{x - 4}$

(E) $\dfrac{dy}{dx} = 2x - y$

10. Evaluate the derivative of $y = \sin x + \cos y$ and solve for $\dfrac{dy}{dx}$.

(A) $\dfrac{dy}{dx} = \dfrac{\sin x}{1 + \cos y}$

(B) $\dfrac{dy}{dx} = \dfrac{\cos x}{\sin y}$

(C) $\dfrac{dy}{dx} = \dfrac{\cos x}{1 - \sin y}$

(D) $\dfrac{dy}{dx} = \dfrac{1 + \cos x}{\sin y}$

(E) $\dfrac{dy}{dx} = \dfrac{\cos x}{1 + \sin y}$

Answers and Explanations

1. C

The easiest way to approach this problem is to write the definition of the derivative of cos $2x$ as the limit of the difference quotient and see if the expression matches one of the choices or is equivalent to one of the choices. We write the definition:

$$\frac{d}{dx}f(x) = \lim_{h \to 0} \frac{\cos 2(x+h) - \cos 2x}{h}$$

This matches (C).

2. B

We follow the procedure for computing the derivative of a polynomial:

$$\frac{d}{dx}\left(x^5 - \sqrt{x} + \frac{1}{x^3}\right) = \frac{d}{dx}\left(x^5\right) - \frac{d}{dx}\left(x^{\frac{1}{2}}\right) + \frac{d}{dx}\left(x^{-3}\right)$$

$$= 5\left(x^{5-1}\right) - \frac{1}{2}\left(x^{-\frac{1}{2}}\right) + (-3)\left(x^{-3-1}\right)$$

$$= 5x^4 - \frac{1}{2}x^{-\frac{1}{2}} - 3x^{-4}$$

$$= 5x^4 - \frac{1}{2\sqrt{x}} - \frac{3}{x^4}$$

3. A

Because the function is a product of two polynomials, we use the Product Rule:

$$\frac{d}{dx}(f(x) \cdot g(x)) = \frac{d}{dx}(f(x)) \cdot g(x) + f(x) \cdot \frac{d}{dx}(g(x))$$

$$\frac{d}{dx}\left((x+7)\left(2x^2 - 3x + 5\right)\right) = \frac{d}{dx}(x+7) \cdot \left(2x^2 - 3x + 5\right) + (x+7) \cdot \frac{d}{dx}\left(2x^2 - 3x + 5\right)$$

$$= (1+0)\left(2x^2 - 3x + 5\right) + (x+7)(4x - 3 + 0)$$

$$= \left(2x^2 - 3x + 5\right) + \left(4x^2 - 3x + 28x - 21\right)$$

$$= 2x^2 - 3x + 5 + 4x^2 + 25x - 21$$

$$= 6x^2 + 22x - 16$$

4. D

Because the function is a quotient of two polynomials, we use the Quotient Rule,

$$\frac{d}{dx}\left(\frac{f(x)}{g(x)}\right)=\frac{\frac{d}{dx}(f(x))\cdot g(x)-f(x)\cdot\frac{d}{dx}(g(x))}{(g(x))^2}.$$

$$\frac{d}{dx}\left(\frac{1}{x+2}\right)=\frac{\frac{d}{dx}(1)\cdot(x+2)-(1)\frac{d}{dx}(x+2)}{(x+2)^2}$$

$$=\frac{(0)(x+2)-(1)(1)}{(x+2)^2}$$

$$=\frac{-1}{(x+2)^2}$$

5. E

We use the Product Rule,

$$\frac{d}{dx}(f(x)\cdot g(x))=\frac{d}{dx}(f(x))\cdot g(x)+f(x)\cdot\frac{d}{dx}(g(x))$$

$$\frac{d}{dx}\left(x^3\cot x\right)=\frac{d}{dx}\left(x^3\right)\cdot\cot x+x^3\cdot\frac{d}{dx}(\cot x)$$

$$=3x^2\cot x+x^3\left(-\csc^2 x\right)$$

$$=3x^2\cot x-x^3\csc^2 x$$

6. C

We use the double-angle identity $\sin 2x = 2 \sin x \cos x$ to simplify the function:

$$\frac{d}{dx}\left(\frac{\cos x}{\sin 2x}\right) = \frac{d}{dx}\left(\frac{\cos x}{2 \sin x \cos x}\right) = \frac{d}{dx}\left(\frac{1}{2 \sin x}\right)$$

We simplify further using the identity $\csc x = \dfrac{1}{\sin x}$:

$$\frac{d}{dx}\left(\frac{1}{2 \sin x}\right) = \frac{d}{dx}\left(\frac{1}{2} \cdot \frac{1}{\sin x}\right) = \frac{1}{2}\frac{d}{dx}(\csc x)$$

We've proven that $\dfrac{d}{dx}(\csc x) = -\csc x \cot x$. Therefore,

$$\frac{1}{2} \cdot \frac{d}{dx}(\csc x) = -\frac{1}{2}\csc x \cot x$$

We could also have used the Quotient Rule to arrive at the same answer.

7. A

Using the Chain Rule, we get

$$\frac{d}{dx}\sqrt{5x^2 + 3x - 1} = \frac{d}{dx}\left(5x^2 + 3x - 1\right)^{\frac{1}{2}}$$

$$= \frac{1}{2}\left(5x^2 + 3x - 1\right)^{-\frac{1}{2}}(10x + 3)$$

$$= \frac{10x + 3}{2\sqrt{5x^2 + 3x - 1}}$$

8. B

We need to use the chain rule because of the composite function $(\tan(\sec x))$.

$$\frac{d}{dx}(\tan(\sec x)) = \sec^2(\sec x) \cdot \sec x \tan x$$

$$= \sec x \tan x \left(\sec^2(\sec x)\right)$$

9. D

We differentiate $x^2 + 4y = xy$ implicitly and solve for $\dfrac{dy}{dx}$:

$$\frac{d}{dx}\left(x^2 + 4y\right) = \frac{d}{dx}(xy)$$

$$2x + 4\frac{dy}{dx} = x \cdot \frac{dy}{dx} + 1 \cdot y$$

$$\frac{dy}{dx} = \frac{2x - y}{x - 4}$$

10. E

First, we differentiate implicitly:

$$y = \sin x + \cos y$$

$$\frac{dy}{dx} = \frac{d}{dx}(\sin x) + \frac{d}{dx}(\cos y) \cdot \frac{dy}{dx}$$

$$= \cos x + (-\sin y)\frac{dy}{dx}$$

$$= \cos x - \sin y \frac{dy}{dx}$$

Now we can solve for $\dfrac{dy}{dx}$:

$$\frac{dy}{dx} + (\sin y)\frac{dy}{dx} = \cos x$$

$$\frac{dy}{dx}(1 + \sin y) = \cos x$$

$$\frac{dy}{dx} = \frac{\cos x}{1 + \sin y}$$

Applications of Derivatives

WHAT ARE THE APPLICATIONS OF DERIVATIVES?

In Chapters 4 and 5, we learned how to graph functions and linear equations using a table of values. In Chapter 6, we analyzed the continuity of graphs and the locations, if any, of vertical asymptotes. In this chapter, we will use the first and second derivatives to analyze a curve. In Chapters 11 and 12, we will explore applications of derivatives involving motion and word problems.

CONCEPTS TO HELP YOU

1. First derivative: The first derivative is the derivative of a function $f(x)$ and is written as $f'(x)$ or $\frac{d}{dx}(f(x))$ or y'. It denotes the slope of $f(x)$ at a point.

2. Second derivative: The second derivative is the derivative of $f'(x)$ and is written as $f''(x)$ or $\frac{d^2}{dx^2}(f(x))$ or y''. It denotes the slope of $f'(x)$ at a point.

3. The sign of $f'(x)$: If $f(x)$ is defined and continuous, then

    • $f(x)$ is increasing on intervals where $f'(x) > 0$

    • $f(x)$ is decreasing on intervals where $f'(x) < 0$

    • $f(x)$ is constant on intervals where $f'(x) = 0$

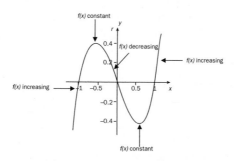

4. The sign of $f''(x)$: Assuming that $f(x)$ is twice differentiable (meaning that both $f'(x)$ and $f''(x)$ exist) on a given interval, then

- if $f''(x) > 0$ on an interval, then $f(x)$ is concave up on that interval
- if $f''(x) < 0$ on an interval, then $f(x)$ is concave down on that interval

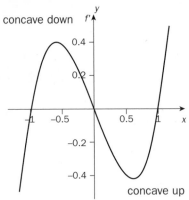

STEPS YOU NEED TO REMEMBER

1. Find the critical points.

Find the critical points of a function by setting the derivative equal to zero and solving for x.

2. Find the relative maxima and minima.

Method 1: Analyze the sign of $f'(x)$ at intervals before and after each critical point to determine where the slope of the original function changes direction. This method is called the First Derivative Test.

Method 2: Analyze the sign of $f''(x)$ at the critical points to determine where the original function is concave up or concave down. This method is called the Second Derivative Test.

3. Find the inflection points.

Set $f''(x) = 0$ and solve for x. Analyze the sign of $f''(x)$ at intervals before and after x to determine where the original function changes concavity.

COMMON APPLICATIONS OF DERIVATIVES QUESTIONS

Equation of the Line Tangent to the Curve at a Point: Write the equations of the tangent lines to the following functions at the given points:

(a) $f(x) = x^2 - 5$ at the point $(1, -4)$ (b) $f(x) = -\frac{1}{2}x^3 + 3$ at $(-2, 7)$

(c) $f(x) = \sin x$ at the point $\left(\frac{\pi}{2}, 1\right)$

Step 1: Compute the derivative of the function.

Step 2: Find the slope of the tangent line by plugging the value of x into the derivative. We will call the slope "m."

Step 3: Substitute the values of the slope and the (x, y) coordinates into the point-slope equation $y - y_0 = m(x - x_0)$.

Step 4: Rearrange the point-slope equation into the slope-intercept equation, $y = mx + b$.

Solution and Explanation: We learned in Chapter 7 that for Δx close to zero, the slope of the line tangent to the graph of $f(x)$ at x_0 is the limit of the slopes of the secant lines between x_0 and $x_0 + \Delta x$.

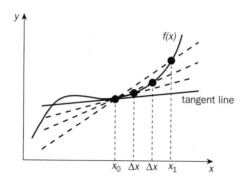

This slope is given by the limit of the difference quotient, which is the derivative at x_0.

(a) The slope of a tangent line to the curve $f(x) = x^2 - 5$ is the derivative of $f(x) = x^2 - 5$: $f'(x^2 - 5) = f'(x^2) - f'(5) = 2x - 0 = 2x$.

At the point $(1,-4)$, the slope is $2(1) = 2$.

Because we know the slope of the line and a point on the line, we can write our equation:
$$y - y_0 = m(x - x_0)$$
$$y - (-4) = 2(x - 1)$$
$$y + 4 = 2x - 2$$
$$y = 2x - 6$$

Our curve and its tangent line at the point $(1,-4)$ look like this:

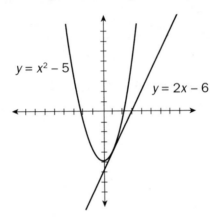

(b) The slope of a tangent line to the curve $f(x) = -\frac{1}{2}x^3 + 3$ is the derivative of $f(x) = -\frac{1}{2}x^3 + 3$.

$$f'\left(-\frac{1}{2}x^3 + 3\right) = f'\left(-\frac{1}{2}x^3\right) + f'(3)$$
$$= -\frac{3}{2}x^2$$

At the point $(-2,7)$, the slope is $-\frac{3}{2}(-2)^2 = -6$.

The equation of the tangent line is

$$y - y_0 = m(x - x_0)$$
$$y - 7 = -6(x + 2)$$
$$y - 7 = -6x - 12$$
$$y = -6x - 5$$

Our curve and its tangent line at the point $(-2, 7)$ look like this:

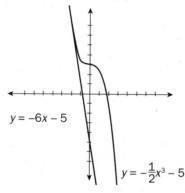

$y = -6x - 5$

$y = -\frac{1}{2}x^3 - 5$

(c) The slope of the tangent line to the curve $f(x) = \sin x$ is $f'(\sin x) = \cos x$.

At the point $\left(\frac{\pi}{2}, 1\right)$, the slope is $\cos \frac{\pi}{2} = 0$.

The equation of the tangent line is

$$y - y_0 = m(x - x_0)$$
$$y - 1 = 0\left(x - \frac{\pi}{2}\right)$$
$$y = 1$$

Our curve and its tangent line at the point $\left(\frac{\pi}{2}, 1\right)$ look like this:

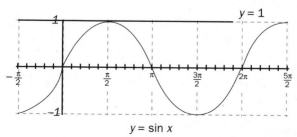

$y = \sin x$

If we examine the graphs of the functions, we find that when the graph is going up, or increasing, the slope of the tangent line is positive, as in example (a), above. When the graph is going down, or decreasing, the slope of the tangent line is negative, as in example (b). When the graph is constant, neither going up or down, the slope of the tangent line is zero, as in example (c).

Points on the graph of a function where the slope of the tangent line is zero or undefined are called the critical points of the function.

Critical Points: Find all critical points of the following functions:

(a) $f(x) = x^4 - 8x^2$ (b) $f(x) = \cos x$ on the interval $[0, 2\pi]$

Step 1: Find the derivative of the function.

Step 2: Set the derivative equal to zero.

Step 3: Solve for x.

Solution and Explanation:

(a) The slope of the tangent line to the curve $f(x) = x^4 - 8x^2$ is $f'(x) = 4x^3 - 16x$. When the slope is zero, the values of x are:

$$4x^3 - 16x = 0$$

$$4x\left(x^2 - 4\right) = 0$$

$$4x(x+2)(x-2) = 0$$

$$4x = 0 \quad \text{or} \quad x+2 = 0 \quad \text{or} \quad x-2 = 0$$
$$x = 0 \qquad\qquad x = -2 \qquad\qquad x = 2$$

The critical points are $x = 0$ or $x = -2$ or $x = 2$.

(b) The slope of the tangent line to the curve $f(x) = \cos x$ is $f'(x) = -\sin x$. When the slope is zero, the values of x on the interval $[0, 2\pi]$ are:

$$-\sin x = 0$$

$$\sin x = 0$$

$$x = 0 \quad \text{or} \quad x = \pi$$

The critical points are $x = 0$ or $x = \pi$.

First Derivative Test: Use the First Derivative Test to find the relative extrema of the following functions:

(a) $f(x) = x^3$ (b) $f(x) = x^2(x-3)$

Step 1: Find the critical points of the function.

Step 2: Do a sign analysis and make a sign graph for $f'(x)$ and indicate where the original function is increasing and decreasing. Label each line.

Step 3: Find the coordinates of all relative maxima.

Step 4: Find the coordinates of all relative minima.

> **Solution and Explanation:** The point where a function changes from increasing to decreasing is its relative maximum; the point where a function changes from decreasing to increasing is its relative minimum. The relative maximum and relative minimum are collectively known as the relative extrema.

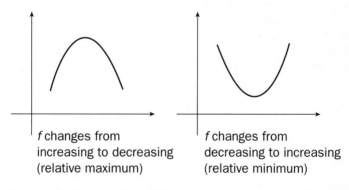

f changes from
increasing to decreasing
(relative maximum)

f changes from
decreasing to increasing
(relative minimum)

The set of points at which $f(x)$ has relative extrema is a subset of the set of critical points.

If we cannot examine the graph of a function, we can use the First Derivative Test to determine whether a critical point is a relative maximum, relative minimum, or neither.

(a) The critical points of $f(x) = x^3$ are:

$$f'\left(x^3\right) = 3x^2$$
$$3x^2 = 0$$
$$x = 0$$

There is only one critical point for $f(x) = x^3$. To see if it is a relative maximum, relative minimum, or neither, we do a sign analysis.

The sign analysis tells us whether or not the function changes direction at the critical point. If the critical point is a relative maximum, then the slope of the curve to the left of zero would be positive and the slope to the right of zero would be negative. If the critical point is a relative minimum, the slope of the curve would change from negative to positive at zero. If there is no relative extrema, the slope would not change signs.

Interval	$(-\infty, 0)$	$(0, \infty)$
Test Point	$x = -1$	$x = 1$
Sign of $f'(x)$	$3x^2 = 3(-1)^2 = 3$	$3x^2 = 3(1)^2 = 3$
Conclusion	$f(x)$ increasing	$f(x)$ increasing

The interval $(-\infty, 0)$ represents the curve to the left of zero and the interval $(0, \infty)$ represents the curve to the right of zero. Since -1 falls within the interval $(-\infty, 0)$ and 1 falls within the interval $(0, \infty)$, we plug these two values into the first derivative to determine whether the slope is positive or negative on either side of zero. We find that $f(x)$ is increasing on both sides of the critical point since the slope of the tangent lines are both positive.

We can see this by looking at the graph of $f(x) = x^3$:

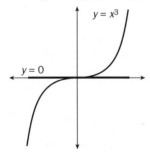

The function is always increasing, so there is no relative extrema. Nevertheless, $x = 0$ is still an important point because the graph has a horizontal tangent there.

Remember: if a function's derivative does not change sign, then it has no relative extrema.

(b) To find the critical points of $f(x) = x^2(x - 3)$, we first expand the function:

$$f(x) = x^2(x - 3) = x^3 - 3x^2$$

Then,

$$f'(x) = 3x^2 - 6x$$

$$3x^2 - 6x = 0$$

$$3x(x - 2) = 0$$

$$3x = 0 \quad \text{or} \quad x - 2 = 0$$

$$x = 0 \qquad\qquad x = 2$$

Now, we do a sign analysis:

Interval	$(-\infty, 0)$	$(0, 2)$	$(2, \infty)$
Test Point	$x = -1$	$x = 1$	$x = 4$
Sign of f'	$f'(-1) = (-3)(-3)$ $= 9$ Positive	$f'(1) = (3)(-2)$ $= -6$ Negative	$f'(4) = (12)(1)$ $= 12$ Positive
Conclusion	f increasing	f decreasing	f increasing

A relative maximum occurs at $x = 0$ because the first derivative changes sign from positive to negative.

A relative minimum occurs at $x = 2$ because the first derivative changes sign from negative to positive.

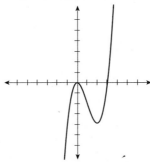

The second derivative can also be used to provide information about a function's critical points. While the first derivative tells us whether a function is increasing or decreasing close to the critical point, the second derivative tells us whether the function is concave up or concave down at the critical point.

Second Derivative Test: If $y = f(x) = x + 2\sin x$, find the critical points of $f(x)$ on the interval $[0, 2\pi]$ and use the Second Derivative Test to determine whether each is a relative maximum or relative minimum.

Step 1: Find the critical points of $f(x)$.

Step 2: Compute the derivative of $f'(x)$ to get $f''(x)$.

Step 3: Plug the critical points into $f''(x)$.

Step 4: Do a sign analysis of $f''(x)$ at the critical points.

> **Solution and Explanation:** The relationship between concavity and the sign of $f''(x)$ is:
>
> - If $f''(x) > 0$ on an interval (i.e., $f''(x)$ is positive), then $f(x)$ is concave up on that interval.
>
> - If $f''(x) < 0$ on an interval (i.e., $f''(x)$ is negative), then $f(x)$ is concave down on that interval.
>
> When we want to determine the behavior of a function at a critical point, it is often easier to use the Second Derivative Test, when it applies (i.e., in places where the function is twice differentiable), than the first derivative test because the sign analysis of $f''(x)$ does not involve creating a table. Instead, it is simply this:
>
> Suppose $f'(x) = 0$. Then:
>
> > If $f''(x) > 0$, then $f(x)$ has a relative minimum at $x = c$.
> >
> > If $f''(x) < 0$, then $f(x)$ has a relative maximum at $x = c$.

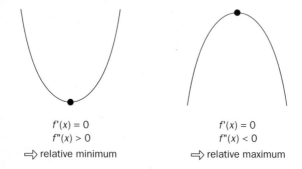

Remember that the critical points of $f(x)$ occur where $f'(x) = 0$ or does not exist. We first compute the derivative of $f(x)$:

$$f(x) = x + 2\sin x$$
$$f'(x) = 1 + 2\cos x$$

The derivative is defined everywhere, so the only critical points occur where $f'(x) = 0$.

$$1 + 2\cos x = 0$$
$$2\cos x = -1$$
$$\cos x = \frac{-1}{2}$$

On the interval $[0, 2\pi]$, this occurs where $x = \frac{2\pi}{3}, \frac{4\pi}{3}$.

We now use the Second Derivative Test to determine the nature of each of these critical points. To apply the test, we compute the second derivative of $f(x)$ and evaluate the second derivative at each of the critical points.

$$f'(x) = 1 + 2\cos x$$
$$f''(x) = -2\sin x$$

For $x = \frac{2\pi}{3}$:

$$f''\left(\frac{2\pi}{3}\right) = -2\sin\left(\frac{2\pi}{3}\right)$$
$$= (-2)\left(\frac{\sqrt{3}}{2}\right)$$
$$= -\sqrt{3}$$

Because $f''\left(\frac{2\pi}{3}\right) < 0$, the Second Derivative Test tells us the $f(x)$ has a relative maximum at $x = \frac{2\pi}{3}$.

For $x = \frac{4\pi}{3}$:

$$f''\left(\frac{4\pi}{3}\right) = -2\sin\left(\frac{4\pi}{3}\right)$$
$$= (-2)\left(-\frac{\sqrt{3}}{2}\right)$$
$$= \sqrt{3}$$

In this case, $f''\left(\dfrac{4\pi}{3}\right) > 0$, so the Second Derivative Test tells us that $f(x)$ has a relative minimum at $x = \dfrac{4\pi}{3}$.

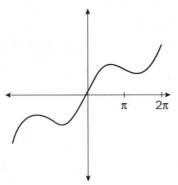

The points at which a function changes concavity are of special interest. These points are called inflection points.

Inflection Points: Find the inflection points of $f(x) = x + 2\sin x$ on the interval $(0, 2\pi)$.

Step 1: Set $f''(x)$ equal to zero.

Step 2: Solve for x on the given interval.

Step 3: Do a sign analysis.

> **Solution and Explanation:** The graphs below show different ways in which a function might change concavity.

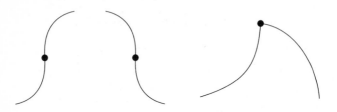

We do a sign analysis, similar to the first derivative sign analysis, to determine whether a point at which the second derivative is zero or undefined is an inflection point.

To find the inflection points of $f(x)$ on $(0,2\pi)$, we first identify the points on $(0,2\pi)$ where $f''(x)$ is zero or undefined. In the previous question, above, we found that $f''(x) = -2\sin x$. This function is defined everywhere, so inflection points can only occur at points where the second derivative is zero. We solve the equation $f''(x) = 0$ to find these points:

$$-2\sin x = 0$$
$$\sin x = 0$$

On the interval $(0,2\pi)$, this occurs where $x = \pi$.

We do a sign analysis:

Interval	$0 < x < \pi$	$\pi < x < 2\pi$
Test Point	$x = \dfrac{\pi}{2}$	$x = \dfrac{3\pi}{2}$
Sign of f''	$f''\left(\dfrac{\pi}{2}\right) = -2\sin\dfrac{\pi}{2} = -2$ negative	$f''\left(\dfrac{3\pi}{2}\right) = -2\sin\dfrac{3\pi}{2} = 2$ positive
Concavity of a	down	up

Because the sign of $f''(x)$ changes at $x = \pi$, $f(x)$ has an inflection point at $x = \pi$.

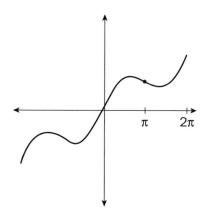

CHAPTER QUIZ

1. Suppose $f(x)$ is a differentiable function with $f(1) = 2$, $f(2) = -2$, $f'(2) = 5$, $f'(1) = 3$, and $f(5) = 1$. An equation of a line tangent to the graph of f is

    (A) $y - 3 = 2(x - 1)$.

    (B) $y - 2 = (x - 1)$.

    (C) $y - 3 = 5(x - 1)$.

    (D) $y - 2 = 3(x - 1)$.

    (E) $y - 1 = 5(x - 2)$.

2. The graph of the differentiable function $y = f(x)$ is shown below. Which of the following is true?

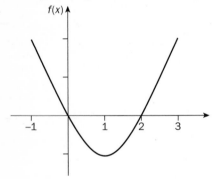

    (A) $f'(0) > f(0)$

    (B) $f'(1) < f(1)$

    (C) $f'(2) < f(2)$

    (D) $f'(1) = f(0)$

    (E) $f'(2) = f(2)$

3. The graph of $y = f(x)$ is shown below. Which of the following graphs could be the derivative?

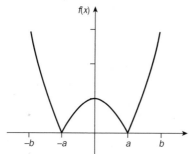

(A)

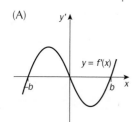

(B)

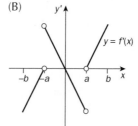

(C)

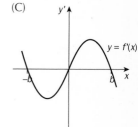

(D)

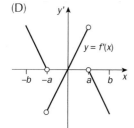

(E)

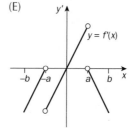

4. Use the First Derivative Test to find the relative extrema of the function $f(x) = \dfrac{x^2 + 2x + 1}{x - 5}$.

   (A) relative maximum at $x = -1$; relative minimum at $x = 11$

   (B) relative maximum at $x = -1$; relative minimum at $x = 5$

   (C) relative maximum at $x = 11$; relative minimum at $x = -1$

   (D) relative maximum at $x = 5$; relative minimum at $x = -1$

   (E) There are no relative extrema.

5. The graph of $f'(x)$ is shown below.

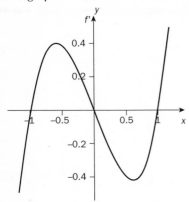

Which of the following graphs could be the graph of $f(x)$?

(A)
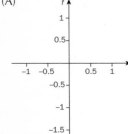

(B)
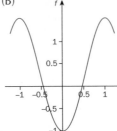

(C)
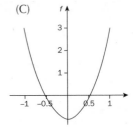

(D)
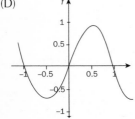

(E)
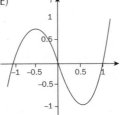

6. Use the Second Derivative Test to find the relative extrema of $f(x) = x^3 - 6x^2 - 36x + 2$.

   (A) relative minimum at $x = 2$, relative maximum at $x = 6$

   (B) relative minimum at $x = -2$, relative maximum at $x = 6$

   (C) relative minimum at $x = 6$, relative maximum at $x = -2$

   (D) relative minimum at $x = -6$, relative maximum at $x = -2$

   (E) None of the above.

7. Given the chart below for the first and second derivatives and assuming that the set $\{a,b,c,d,e,f,g\}$ contains all the critical points, at which value(s) of x does a relative minimum occur?

x	a	b	c	d	e	f	g
$f'(x)$	negative	zero	positive	zero	negative	infinite	positive
$f''(x)$	positive	positive	zero	negative	negative	infinite	negative

   (A) a

   (B) b

   (C) c

   (D) d

   (E) e

8. Find the point of inflection of $g(x) = x^2 - \dfrac{8}{x}$, $x > 0$.

   (A)   1

   (B)   2

   (C)   4

   (D)   8

   (E)   16

9. If $s''(t) = (t+1)(t-3)\sin^2 t$, then the inflection point(s) of the function $s(t)$ is/are at

(A) $-1, 3$.

(B) -1 only.

(C) 3 only.

(D) $n\pi$, where n is an integer.

(E) $-1, 3, n\pi$, where n is an integer.

10. The graph of the derivative of $y = f(x)$ on the interval $x_0 \le x \le x_4$ is shown below.

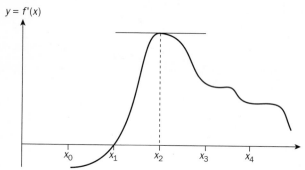

$y = f'(x)$

Which of the following lists all the points of inflection?

(A) x_1

(B) x_2

(C) x_3

(D) x_1, x_3

(E) x_2, x_3

Answers and Explanations

1. D

The equation of the line tangent to the graph of $f(x)$ is given by $y - y_0 = m(x - x_0)$.

We are given information about three values of x: $x = 1$, $x = 2$, and $x = 5$. To find the equation of the line tangent to the curve at a given point, we must know both the value of the *function* and the value of the *derivative* at the given point. This is clear if we organize the given information in a table:

x	$f(x)$	$f'(x)$	$y - y_0 = m(x - x_0)$
1	2	3	$y - 2 = 3(x - 1)$
2	−2	5	$y + 2 = 5(x - 2)$
5	1	?	not enough information

Now we can easily match the information on the table with the options. Only one matches the equation of a tangent line on the table—(D).

2. D

(A) False: $f'(0) < 0$ and $f(0) = 0$

(B) False: $f'(1) = 0$ and $f(1) < 0$

(C) False: $f'(2) > 0$ and $f(2) = 0$

(D) True: $f'(1) = 0$ and $f(0) = 0$

(E) False: $f'(2) > 0$ and $f(2) = 0$

3. B

Let's describe the graph in terms of its derivative, because this will allow us to eliminate various choices.

On $x < -a$, the slope of the curve is negative, so the graph of y' lies below the x-axis. Thus choices (A) and (D) are eliminated. On $-a < x < 0$, the slope of the curve is positive, so the graph of y' lies above the x-axis. This means that (C), (D), and (E) are not possible answers. We've eliminated all of the choices except one, so the correct answer is (B).

Notice that the graph of $f(x)$ has sharp corners at $x = -a$ and $x = a$. This means that the function is *not* differentiable at those two points. Therefore, the graph of the derivative has holes at $-a$ and a.

4. A

We use the Quotient Rule to find the derivative of $f(x) = \dfrac{x^2 + 2x + 1}{x - 5}$:

$$f'(x) = \frac{(2x+2)(x-5) - \left(x^2 + 2x + 1\right)(1)}{(x-5)^2}$$

$$= \frac{2x^2 - 8x - 10 - x^2 - 2x - 1}{(x-5)^2}$$

$$= \frac{x^2 - 10x - 11}{(x-5)^2}$$

To find the critical points, we set the derivative equal to zero:

$$\frac{x^2 - 10x - 11}{(x-5)^2} = 0$$

$$x^2 - 10x - 11 = 0$$

$$(x-11)(x+1) = 0$$

$$
\begin{array}{ccccc}
x - 11 = 0 & & x + 1 = 0 & & x - 5 = 0 \\
& \text{or} & & \text{or} & \\
x = 11 & & x = -1 & & x = 5
\end{array}
$$

Because the derivative is undefined when the denominator is equal to zero, $x = 5$ is also a critical point.

Now, we do a sign analysis to see if any of the critical points are a relative extrema.

Interval	$(-\infty, -1)$	$(-1, 5)$	$(5, 11)$	$(11, \infty)$
Test Point	-2	0	6	12
Sign of $f'(x)$	$f'(-2) = \dfrac{26}{49}$ positive	$f'(0) = -\dfrac{11}{25}$ negative	$f'(6) = -35$ negative	$f'(12) = \dfrac{13}{49}$ positive
Conclusion	$f(x)$ increasing	$f(x)$ decreasing	$f(x)$ decreasing	$f(x)$ increasing

What matters are the signs of the derivatives, so the function has a relative maximum at $x = -1$ because the sign of the derivative changes from positive to negative at that point, and the function has a relative minimum at $x = 11$ because the sign of the derivative changes from negative to positive at that point.

5. A

We can approach this problem by eliminating choices. Interpreting the graph of $f'(x)$, we can see that, by the First Derivative Test,

- $f(x)$ has a relative minimum at $x = -1$; the function changes from decreasing to increasing at this point, because the sign of the first derivative changes from negative to positive.
- $f(x)$ has a relative maximum at $x = 0$; $f(x)$ changes from increasing to decreasing here, because the sign of the first derivative changes from positive to negative.
- $f(x)$ has another relative maximum at $x = 1$; the function changes from decreasing to increasing here, because the sign of the first derivative changes from negative to positive.

Using this information, we can eliminate option (B) because this graph has the maxima and minima reversed. Option (C) can be eliminated because this graph has only one relative extremum. Options (D) and (E) cannot be correct because the relative extrema of these graphs are not located at the correct points. The only remaining choice is (A). We can confirm on the graph that the graph of (A) has relative minima at $x = -1$ and 1 and a relative maximum at $x = 0$.

6. C

The critical points of $f(x) = x^3 - 6x^2 - 36x + 2$ are

$$f'(x) = 3x^2 - 12x - 36$$
$$3x^2 - 12x - 36 = 0$$
$$(3x + 6)(x - 6) = 0$$
$$3x + 6 = 0 \quad \text{or} \quad x - 6 = 0$$
$$x = -2 \qquad\qquad x = 6$$

The second derivative is

$$f'(x) = 3x^2 - 12x - 36$$
$$f''(x) = 6x - 12$$

Plugging the critical points into $f''(x)$, we get

$$f''(-2) = 6(-2) - 12 \quad \text{and} \quad f''(6) = 6(6) - 12$$
$$= -24 \qquad\qquad\qquad = 24$$

Because $f''(-2) < 0$, the function is concave down and has a relative maximum at $x = -2$. The function is concave up and has a relative minimum at $x = 6$ because $f''(6) > 0$.

7. B

Critical points occur where $f'(x) = 0$. According to the chart,

$$f'(x) = 0$$
$$x = b, d$$

We are also given that $f''(b) > 0$ and $f''(d) < 0$; therefore, the function is concave up at $x = b$ and has a relative minimum at $x = b$ only.

8. B

A point of inflection occurs when $g''(x)$ changes sign. We compute:

$$g'(x) = 2x + \frac{8}{x^2}$$
$$g''(x) = 2 - \frac{16}{x^3}$$

We set $g''(x) = 0$ and solve the resulting equation:

$$2 - \frac{16}{x^3} = 0$$
$$-\frac{16}{x^3} = -2$$
$$x^3 = 8$$
$$x = 2$$

We use the point $x = 2$ to set up a sign analysis for the second derivative:

Interval	$0 < x < 2$	$x > 2$
Test Point	$x = 1$	$x = 3$
Sign of g''	$g''(1) = 2 - \dfrac{16}{1} = -14$ negative	$g''(3) = 2 - \dfrac{16}{27} \approx 1\dfrac{1}{2}$ positive
Concavity of g	down	up

Because $g''(x)$ changes from negative to positive at $x = 2$ (i.e., because $g(x)$ changes concavity), $g(x)$ has an inflection point at $x = 2$.

9. A

The only possible points at which the second derivative could change sign are places where it equals zero. We solve the equation $s''(t) = (t + 1)(t - 3)\sin^2 t = 0$ to find $s''(t) = 0$ at

$$
\begin{array}{ccccc}
& & & & \sin^2 t = 0 \\
t + 1 = 0 & & t - 3 = 0 & & \\
& \text{or} & & \text{or} & \sin t = 0 \\
t = -1 & & t = 3 & & \\
& & & & t = 0\pi, 1\pi, 2\pi, \ldots
\end{array}
$$

Notice that the factor $\sin^2 t$ in the equation $s''(t)$ appears to the second power, which means that it is always positive (i.e., $\sin^2 t > 0$). Therefore, the sign of $s''(t)$ does not depend on the sign of this factor and the second derivative will not change sign at $n\pi$ (i.e., 0π, 1π, 2π, etc.). Thus we need only include $t = -1$ and 3 in our sign analysis of the second derivative.

Interval	$t < -1$	$-1 < t < 3$	$t > 3$
Test Point	$t = -2$	$t = 0$	$t = 4$
Sign of s''	$s''(-2) = (-2 + 1)(-2 - 3)(+)$ $= (-)(+)(-) = (+)$ positive	$s''(0) = (0 + 1)(0 - 3)(+)$ $= (+)(+)(-) = (-)$ negative	$s''(4) = (4 + 1)(4 - 3)(+)$ $= (+)(+)(+) = (+)$ positive
Concavity	up	down	up

Because $s''(t)$ changes sign at both $t = 3$ and $t = -1$, both of these are inflection points.

10. B

The second derivative of $f(x)$ is the derivative of $f'(x)$. The derivative of $f'(x)$ at a point is the slope of $f'(x)$ at that point.

Because $f(x)$ is concave up where the second derivative is positive (i.e., the slope of the first derivative is positive), the graph of $f(x)$ is concave up where the graph of $f'(x)$ is going up or where the graph of $f''(x)$ is above the x-axis.

Because $f(x)$ is concave down where the second derivative is negative (i.e., the slope of the first derivative is negative), the graph of $f(x)$ is concave down where the graph of $f'(x)$ is decreasing or where the graph of $f''(x)$ is below the x-axis.

The function $f(x)$ has inflection points where $f''(x)$ changes sign.

We can see on the graph of $f'(x)$ that the sign of the slope of $f'(x)$ changes from positive to negative at x_2— where the graph changes from increasing to decreasing—at no other point, so the correct answer is (B).

CHAPTER 9

Integration

WHAT IS INTEGRATION?

In the last two chapters, we learned about derivatives, which are one of two main subject areas in calculus. In this chapter, we will learn about the other major subject area: integrals. Integration, or antidifferentiation, is the inverse of differentiation. You can use what you've learned about computing derivatives to compute antiderivatives, and your knowledge of antiderivatives will help you to evaluate indefinite and definite integrals.

CONCEPTS TO HELP YOU

1. Antiderivatives: If $\frac{d}{dx}(F(x)) = f(x)$, then $F(x)$ is an antiderivative of $f(x)$.

2. Indefinite integrals: If $F(x)$ is an antiderivative of $f(x)$, the indefinite integral is $\int f(x)dx = F(x) + C$, where there are infinite possibilities for the value of C. Properties of the indefinite integral are:

   • $\int f(x)\ dx \pm \int g(x)\ dx = \int f(x) \pm g(x)\ dx$
   • $\int c \cdot f(x)\ dx = c \int f(x)\ dx$

3. Techniques of antidifferentiation: Integration formulas are derived from their corresponding derivative formulas.

Because the derivative is...	The indefinite integral is...
$\dfrac{d}{dx}(x) = 1$	$\displaystyle\int 1\ dx = x + C$
$\dfrac{d}{dx}\left(\dfrac{x^{r+1}}{r+1}\right) = x^r,\ r \neq -1$	$\displaystyle\int x^r\ dx = \dfrac{x^{r+1}}{r+1} + C$
$\dfrac{d}{dx}(\sin x) = \cos x$ $\quad$ $\dfrac{d}{dx}(-\cos x) = \sin x$ $\dfrac{d}{dx}(\tan x) = \sec^2 x$ $\quad$ $\dfrac{d}{dx}(-\cot) = \csc^2 x$ $\dfrac{d}{dx}(\sec x) = \sec x \tan x$ $\quad$ $\dfrac{d}{dx}(-\sec x) = \csc x \cot x$	$\displaystyle\int \cos x\ dx = \sin x + C$ $\quad$ $\displaystyle\int \sin x\ dx = -\cos x + C$ $\displaystyle\int \sec^2 x\ dx = \tan x + C$ $\quad$ $\displaystyle\int \csc^2 x\ dx = -\cot x + C$ $\displaystyle\int \sec x \tan x\ dx = \sec x + C$ $\quad$ $\displaystyle\int \csc x \cot x\ dx = -\csc x + C$
$\dfrac{d}{dx}\left(e^x\right) = e^x$	$\displaystyle\int e^x\ dx = e^x + C$
$\dfrac{d}{dx}(\ln x) = \dfrac{1}{x}$	$\displaystyle\int \dfrac{1}{x}\ dx = \ln|x| + C$
$\dfrac{d}{dx}\left(\dfrac{b^x}{\ln b}\right) = b^x$ for $b > 0$	$\displaystyle\int b^x\ dx = \dfrac{b^x}{\ln b} + C$
$\dfrac{d}{dx}(\arctan x) = \dfrac{1}{1+x^2}$ $\dfrac{d}{dx}(\arcsin x) = \dfrac{1}{\sqrt{1-x^2}}$ $\dfrac{d}{dx}(\text{arcsec } x) = \dfrac{1}{|x|\sqrt{x^2-1}}$	$\displaystyle\int \dfrac{1}{1+x^2}\ dx = \arctan x + C$ $\displaystyle\int \dfrac{1}{\sqrt{1-x^2}}\ dx = \arcsin x + C$ $\displaystyle\int \dfrac{1}{x\sqrt{x^2-1}}\ dx = \text{arcsec}|x| + C$

4. Definite integrals: The definite integral $\int_a^b f(x)\,dx$ differs from the indefinite integral in that the definite integral has limits of integration, indicating boundaries when calculating the integral. Properties of the definite integral are:

    - $\int_a^b f(x)\,dx \pm \int_a^b g(x)\,dx = \int_a^b f(x) \pm g(x)\,dx$

    - $\int_a^b c \cdot f(x)\,dx = c \int_a^b f(x)\,dx$

5. Fundamental Theorem of Calculus: If $F(x)$ is an antiderivative of the continuous function $f(x)$, then $\int_a^b f(x)\,dx = F(b) - F(a)$.

STEPS YOU NEED TO REMEMBER

1. *Compute the antiderivative.*

Learn the information and relationships in the table of techniques of differentiation, above. Use this information, or the table itself, to compute the antiderivative. Because antiderivatives are the opposite of derivatives, we can also begin by asking ourselves "What function has $f(x)$ as its derivative?" to arrive at the antiderivative.

2. *Include + C in the solution for indefinite integrals.*

If you are solving an indefinite integral, you need to follow up the antiderivative with the expression "+ C."

3. *Use the Fundamental Theorem of Calculus to solve definite integrals.*

Unlike the indefinite integral, the solution for the definite integral is a real number value. We use $\int_a^b f(x)\,dx = F(b) - F(a)$—the formula from the Fundamental Theorem of Calculus—to solve the definite integral.

COMMON INTEGRATION QUESTIONS

Indefinite Integrals: Evaluate the indefinite integrals:

(a) $\int 2x^3 \, dx$ (b) $\int \left(5x^2 + 3x - 6\right) dx$

Step 1: Compute the antiderivative.

Step 2: Take the derivative of the answer to make sure you have computed correctly.

Step 3: Include the constant $+ C$ in the solution.

Solution and Explanation: If $F(x)$ is an antiderivative of $f(x)$, we write $\int f(x) \, dx = F(x) + C$. We call the symbol $\int \ dx$ the *indefinite integral* and we can read the expression $\int f(x) \, dx$ as "the indefinite integral $f(x)$." The function $f(x)$ is called the *integrand* and C is the constant of integration.

(a) Recall from the chapter on derivatives that $\dfrac{d}{dx}\left(x^r\right) = r \cdot x^{r-1}$. It follows, then, that $\dfrac{d}{dx}\left(\dfrac{x^{r+1}}{r+1}\right) = r + 1 \cdot \dfrac{x^{r+1-1}}{r+1} = x^r$. Therefore, an antiderivative of x^r is $\dfrac{x^{r+1}}{r+1}$.

Based on the properties of indefinite integrals, we can rewrite $\int 2x^3 \, dx$ as $2\int x^3 \, dx$.

An antiderivative of x^3 is $\dfrac{x^{3+1}}{3+1} = \dfrac{x^4}{4}$, so an antiderivative of $2x^3$ is $2 \cdot \dfrac{x^4}{4} = \dfrac{x^4}{2}$.

To make sure we computed correctly, we find $\dfrac{d}{dx}\left(\dfrac{x^4}{2}\right) = \dfrac{4x^{4-1}}{2} = 2x^3$.

Notice that we say that $\dfrac{x^4}{2}$ is *an* antiderivative and not *the* antiderivative of $2x^3$. This is because $\dfrac{x^4}{2}$ is not the only antiderivative of $2x^3$. Recall that $\dfrac{d}{dx}(c) = 0$, where c is a constant. Because the derivative of a constant is zero, an antiderivative of $2x^3$ can also be $\dfrac{x^4}{2} + 1$ and $\dfrac{x^4}{2} - 6$:

$$\frac{d}{dx}\left(\frac{x^4}{2}+1\right)=\frac{d}{dx}\left(\frac{x^4}{2}\right)+\frac{d}{dx}(1)=2x^3+0=2x^3$$

and

$$\frac{d}{dx}\left(\frac{x^4}{2}-6\right)=\frac{d}{dx}\left(\frac{x^4}{2}\right)-\frac{d}{dx}(6)=2x^3-0=2x^3.$$

In fact, there are an infinite number of antiderivatives for $2x^3$ and they differ only by the value of the constant. Hence,

$$\int 2x^3\ dx=\frac{x^4}{2}+C$$

(b) Based on the properties of indefinite integrals,

$$\int\left(5x^2+3x-6\right)dx=\int\left(5x^2\right)dx+\int(3x)\ dx-\int(6)\ dx$$
$$=5\int\left(x^2\right)dx+3\int(x)\ dx-\int(6)\ dx$$
$$=5\cdot\frac{x^3}{3}+3\cdot\frac{x^2}{2}-6\cdot\frac{x^1}{1}+C$$
$$=\frac{5x^3}{3}+\frac{3x^2}{2}-6x+C$$

When the integrand is complicated, we can use the substitution method to simplify it into something we can integrate directly.

Substitution Method: Compute by substitution:

(a) $\int 2x\cos x^2\ dx$ (b) $\int xe^{x^2}\ dx$

Step 1: Substitute the integrand with u and du.

Step 2: Compute the antiderivative in terms of u.

Step 3: Recover the x by substituting back into the antiderivative.

Solution and Explanation: The key to u-substitution is to find a function in the integrand whose derivative is also part of the integrand.

In the indefinite integral $\int 2x\cos x^2\ dx$, we see that the derivative of x^2 (in $\cos x^2$) is $2x$, which is also part of the integrand.

Therefore, we set $u = x^2$ and $du = 2x\,dx$ and rewrite the indefinite integral as

$$\int 2x\cos x^2 \; dx = \int \cos \underbrace{x^2}_{u} \cdot \underbrace{2x \; dx}_{du} = \int \cos u \; du.$$

The integrand written in terms of u is something we can integrate directly, using the table on differentiation or asking ourselves "What function has $\cos u$ as its derivative?" The answer is:

$$\int \cos u \; du = \sin u + C$$

To recover the x, we substitute back:

$$\int \cos u \; du = \sin u + C = \sin x^2 + C.$$

Therefore, $\int 2x \cos x^2 \; dx = \sin x^2 + C$

(b) Computing $\int xe^{x^2} \, dx$ directly is tricky. We can try to make the u-substitution $u = x^2$ and $du = 2x\,dx$. However, we don't have a $2x$ in the integral—we only have an x.

Fortunately, we can manipulate *constants*:

$$u = x^2$$
$$du = 2x \; dx$$
$$\frac{1}{2} du = x \; dx$$

Therefore:

$$\int xe^{x^2} \, dx = \int e^{\overset{u}{\overbrace{x^2}}} \cdot \underbrace{x \; dx}_{\frac{1}{2} du}$$

$$= \int e^u \cdot \frac{1}{2} du$$

$$= \frac{1}{2} \int e^u \, du$$

$$= \frac{1}{2} e^u + C$$

Substituting back we get:

$$\frac{1}{2}e^{u} + C = \frac{1}{2}e^{x^2} + C$$

Therefore, $\int xe^{x^2}\, dx = \frac{1}{2}e^{x^2} + C$.

If the integrand contains two functions not related to one another, we can use integration by parts.

Integration by Parts Method: Compute by integration by parts:

(a) $\int xe^{x}\, dx$ (b) $\int \ln x\; dx$

Step 1: Substitute the integrand with u and dv.

Step 2: Use the integration by parts formula $\int u\; dv = uv - \int v\; du$.

Step 3: Compute $\int v\; du$.

Step 4: Simplify.

> **Solution and Explanation:** The integration by parts formula is $\int u\; dv = uv - \int v\; du$. To use the integration by parts formula, we need to decide which function is u and which function is dv. Much like the substitution method, making this choice properly really just takes practice. If the formula doesn't work the way you think it should with your initial guess, then just guess again. In general, however, it helps to keep in mind that we would like u to be simpler after *differentiating*, while we want dv to be simpler after *integrating*.
>
> (a) In $\int xe^{x}\, dx$ we have two functions. We cannot use u-substitution because the derivative of x (in e^x) is 1. In problem 2(b), above, we could manipulate *constants* so that we were able to perform a substitution, but we cannot manipulate anything else. In other words:
>
> $u = x$
> $du \neq dx$
> $x\,du = x\,dx$

To use integration by parts, let $u = x$ and $dv = e^x \, dx$. Remember that we want u to be simpler after *differentiating* and we want dv to be simpler after *integrating*:

$$u = x \qquad dv = e^x \, dx$$
$$\text{and}$$
$$du = dx \qquad v = \int dv = \int e^x \, dx = e^x$$

Then using the integration by parts formula, we have

$$\int x e^x \, dx = x e^x - \int e^x \, dx$$
$$= x e^x - e^x$$
$$= e^x (x - 1) + C$$

and we're done.

(b) At first glance, $\int \ln x \, dx$ may not appear to contain two functions. To help you see that there actually *are* two functions in the integrand, let's write it as $\int \ln x \cdot 1 dx$.

To solve using integration by parts, we let

$$u = \ln x \qquad dv = dx$$
$$\text{and}$$
$$du = \frac{1}{x} dx \qquad v = x$$

Therefore, we get:

$$\int \ln x \, dx = x \ln x - \int x \cdot \frac{1}{x} dx$$
$$= x \ln x - \int dx$$
$$= x \ln x - x + C$$

If we apply an interval—the limits of integration—to an indefinite integral, we get a definite integral. The Fundamental Theorem of Calculus allows us to compute the definite integral to get a real number solution.

Definite Integrals: Evaluate the definite integrals.

(a) $\int\limits_{1}^{4} 2x^3 \ dx$ (b) $\int\limits_{2}^{5}\left(x^2 - 4x + 1\right)dx$

Step 1: Compute the antiderivative of the function at the upper limit of integration.

Step 2: Compute the antiderivative of the function at the lower limit of integration.

Step 3: Subtract the two antiderivatives.

Solution and Explanation: The Fundamental Theorem of Calculus states that if f is continuous on $[a,b]$ and F is any antiderivative of f, then $\int\limits_{a}^{b} f(x)dx = F(b) - F(a)$. We often use the shorthand "FTC" instead of writing "Fundamental Theorem of Calculus" and we sometimes use the notation $F(x)\big|_{a}^{b}$ or $F(x)\big]_{a}^{b}$ instead of writing $F(b) - F(a)$.

Unlike the indefinite integral, the solution to a definite integral is a real number because the constant of integration ($+ C$) cancels out:

$$\int\limits_{a}^{b} f(x)dx = \left[F(b) + C\right] - \left[F(a) + C\right]$$

$$= F(b) + C - F(a) - C$$

$$= F(b) - F(a)$$

(a) We know from the previous question that $\int 2x^3 \ dx = \dfrac{x^4}{2} + C$, where $\int 2x^3 dx$ is the *indefinite integral* and $\dfrac{x^4}{2}$ is an *antiderivative* of $2x^3$. In this problem, we have placed limits of integration on the function $2x^3$, so we can calculate the *definite integral*.

Using the FTC, we get:

$$\int_1^4 2x^3\,dx = \frac{x^4}{2}\bigg|_1^4$$

$$= \frac{(4)^4}{2} - \frac{(1)^4}{2}$$

$$= \frac{256}{2} - \frac{1}{2}$$

$$= \frac{255}{2} = 127.5$$

(b) Based on the properties of definite integrals, we can rewrite

$$\int_2^5 \left(x^2 - 4x + 1\right)dx \text{ as } \int_2^5 x^2\,dx - 4\int_2^5 x\,dx + \int_2^5 1\,dx.$$

The antiderivative of x^2 is $\dfrac{x^{2+1}}{2+1} = \dfrac{1}{3}x^3$.

The antiderivative of x is $\dfrac{x^{1+1}}{1+1} = \dfrac{1}{2}x^2$.

The antiderivative of 1 (i.e., x^0) is $\dfrac{x^{0+1}}{0+1} = x$.

Therefore,

$$\int_2^5 \left(x^2 - 4x + 1\right)dx = \int_2^5 x^2\,dx - 4\int_2^5 x\,dx + \int_2^5 1\,dx$$

$$= \left[\frac{1}{3}x^3 - 4\cdot\frac{1}{2}x^2 + x\right]_2^5$$

$$= \left[\frac{1}{3}(5)^3 - 2(5)^2 + 5\right] - \left[\frac{1}{3}(2)^3 - 2(2)^2 + 2\right]$$

$$= \left(-\frac{10}{3}\right) - \left(-\frac{10}{3}\right)$$

$$= 0$$

For more complicated definite integrals, we can use the substitution and integration by parts techniques that we used to solve indefinite integrals.

Substitution and Integration by Parts of Definite Integrals: Evaluate the definite integrals using either the substitution or integration by parts method:

(a) $\displaystyle\int_1^2 \frac{x}{\left(x^2+2\right)^3} \, dx$ (b) $\displaystyle\int_{\frac{\pi}{3}}^{\frac{\pi}{2}} x \sin x \, dx$

Step 1: Determine whether to use substitution or integration by parts.

Step 2: Compute the antiderivative $F(x)$.

Step 3: Solve on the limits of integration using the Fundamental Theorem of Calculus.

Solution and Explanation:

(a) We will use the substitution method for $\displaystyle\int_1^2 \frac{x}{\left(x^2+2\right)^3} \, dx$ because the integrand is made up of a function and the derivative (or part of the derivative) of the function:

$$u = x^2 + 2$$
$$du = 2x \, dx$$

Recall that we can manipulate *constants* in the integrand, so we can multiply by $\frac{1}{2}$ to cancel out the constant in $2x$:

$$du = 2x \, dx$$
$$\frac{1}{2} du = x \, dx$$

We also need to convert the limits of integration from their x values to their corresponding u values.

Since $u = x^2 + 2$, the upper limit of integration becomes $u = (2)^2 + 2 = 6$ and the lower limit of integration becomes $u = (1)^2 + 2 = 3$.

Therefore,

$$\int_1^2 \frac{x}{\left(x^2+2\right)^3}\,dx = \int_3^6 \frac{1}{2}\cdot\frac{du}{u^3}$$

$$= \frac{1}{2}\int_3^6 u^{-3}\,dus$$

$$= \frac{1}{2}\left(-\frac{1}{2}\cdot u^{-2}\right)\Big|_3^6$$

$$= -\frac{1}{4}\left[(6)^{-2}-(3)^{-2}\right]$$

$$= -\frac{1}{4}\left(\frac{1}{36}-\frac{1}{9}\right)$$

$$= \frac{1}{48}$$

(b) We will use integration by parts to compute $\int_{\frac{\pi}{3}}^{\frac{\pi}{2}} x\sin x\,dx$ because the

integrand is made up of two functions (x and $\sin x$) that are not related. Let

$$u = x \qquad\qquad dv = \sin x\,dx$$
$$\text{and}$$
$$du = dx \qquad\qquad v = -\cos x$$

Recall that the integration by parts formula is $\int u\,dv = uv - \int v\,du$, so

$$\int x\sin x\,dx = -x\cos x - \int -\cos x\,dx$$

$$= -x\cos x + \sin x + C$$

On the interval $\frac{\pi}{3} \le x\,\frac{\pi}{2}$,

$$\int_{\frac{\pi}{3}}^{\frac{\pi}{2}} x\sin x\,dx = \left[-x\cos x + \sin x\right]_{\frac{\pi}{3}}^{\frac{\pi}{2}}$$

$$= \left[\left(-\frac{\pi}{2}\right)\left(\cos\frac{\pi}{2}\right)+\sin\frac{\pi}{2}\right] - \left[\left(-\frac{\pi}{3}\right)\left(\cos\frac{\pi}{3}\right)+\sin\frac{\pi}{3}\right]$$

$$= (0+1) - \left(-\frac{\pi}{6}+\frac{\sqrt{3}}{2}\right)$$

$$= 1 + \frac{\pi}{6} - \frac{\sqrt{3}}{2}$$

CHAPTER QUIZ

1. $\int \frac{d}{dx}\left(F(x)\right)dx =$

    (A) $F(x)$

    (B) $F'(x)$

    (C) $F'(x) + C$

    (D) $F(x) + C$

    (E) Cannot be determined

2. If $\tan^2 x$ is an antiderivative of $f(x)$, which of the following *cannot* be an antiderivative of $f(x)$?

    (A) $\sec^2 x$

    (B) $\tan^2 x + 3$

    (C) $\frac{\sin^2 x}{\cos^2 x}$

    (D) $\frac{\sin^2 x}{1 - \cos^2 x}$

    (E) $\frac{1}{\cos^2 x}$

3. $\int 2e^x + \frac{5}{x^2}\, dx =$

    (A) $e^{2x} - \frac{5}{x} + C$

    (B) $e^{2x} - \frac{10}{x^3} + C$

    (C) $2e^x - \frac{5}{x} + C$

    (D) $2e^x - \frac{10}{x^3} + C$

    (E) Cannot be determined

4. Evaluate $\int x^2 \left(x^3 + 1\right)^5 dx$.

    (A) $\frac{1}{18}\left(x^3 + 1\right)^6 + C$

    (B) $\frac{1}{6}\left(x^3 + 1\right)^6 + C$

    (C) $\frac{1}{2}\left(x^3 + 1\right)^6 + C$

    (D) $\frac{1}{18}\left(x^3 + 1\right)^5 + C$

    (E) $\frac{1}{18}\left(x^3 + 1\right)^4 + C$

5. Evaluate $\int x \sec^2 x\, dx$.

    (A) $\ln x - \ln|\sec x| + C$

    (B) $x \ln x + \ln|\sec x| + C$

    (C) $x - \ln|\sec x| + C$

    (D) $x \tan x - \ln|\sec x| + C$

    (E) $x \ln x - \ln|\tan x| + C$

6. Suppose $f(x)$ is a continuous function and f is the derivative of the function $F(x)$. This is a table of values for $f(x)$ and $F(x)$:

x	0	1	2	3
$f(x)$	−1	0	1	−2
$f(x)$	4	3	A	8

    What is $\int_1^3 f(x)dx$?

    (A) 5

    (B) 8

    (C) −2

    (D) 19

    (E) Cannot be determined

7. $\displaystyle\int_{\frac{5\pi}{6}}^{\frac{5\pi}{4}} \sec^2\theta \; d\theta =$

   (A) $1 + \dfrac{\sqrt{3}}{3}$

   (B) $1 + \sqrt{3}$

   (C) $1 - \dfrac{\sqrt{3}}{3}$

   (D) $1 - \sqrt{3}$

   (E) $-\dfrac{\sqrt{3}}{3} - 1$

8. $\displaystyle\int_0^{\sqrt{2}} \left(3x - \sqrt{2}\right)^5 dx =$

   (A) 504

   (B) 168

   (C) 84

   (D) 28

   (E) −84

9. $\displaystyle\int_{-1}^{0} \dfrac{8x^3 - 12x^2 + 7x + 2}{2x - 2} \; dx$

   (A) $\dfrac{23}{6}$

   (B) 8

   (C) $\dfrac{23}{6} - \dfrac{5}{2}\ln 2$

   (D) $-\dfrac{5}{2}\ln 2 \; \dfrac{1}{2}\left(2e^3 + 1\right)$

   (E) $8 + \dfrac{5}{2}\ln 2$

10. Evaluate $\displaystyle\int_1^{e} x^2 \ln x \; dx$.

   (A) $\dfrac{1}{6}\left(2e^3 + 1\right)$

   (B) $\dfrac{1}{3}\left(2e^3 + 1\right)$

   (C) $\left(2e^3 + 1\right)$

   (D) $\dfrac{1}{2}\left(2e^3 + 1\right)$

   (E) $\dfrac{1}{9}\left(2e^3 + 1\right)$

Answers and Explanations

1. D

The indefinite integral of a function $\int f(x)\,dx$ is any antiderivative of f plus a constant. An antiderivative of $\dfrac{d}{dx}(F(x))$ is $F(x)$; therefore, $\int \dfrac{d}{dx}(F(x))\,dx = F(x) + C$.

2. D

Two antiderivatives of the same function must differ by a constant. We can use trig identities to assess each option.

(A) $\sec^2 x = \tan^2 x + 1$, is an antiderivative of $f(x)$

(B) $\tan^2 x + 3$ is an antiderivative of $f(x)$

(C) $\dfrac{\sin^2 x}{\cos^2 x} = \tan^2 x$ is an antiderivative of $f(x)$

(D) $\dfrac{\sin^2 x}{1 - \cos^2 x} = \dfrac{\sin^2 x}{\sin^2 x} = 1$

(E) $\dfrac{1}{\cos^2 x} = \sec^2 x = \tan^2 x + 1$ is an antiderivative of $f(x)$

By process of elimination, (D) is correct.

3. C

We use the properties of indefinite integrals to rewrite the given integral as $\int 2e^x + \dfrac{5}{x^2}\,dx = 2\int e^x\,dx + 5\int x^{-2}\,dx$.

The antiderivative of e^x is e^x because $\dfrac{d}{dx}(e^x) = e^x$ and the antiderivative of x^{-2} is $-x^{-1}$ because $\dfrac{d}{dx}(-x^{-1}) = x^{-2}$, so

$$\int 2e^x + \frac{5}{x^2}\,dx = 2\int e^x\,dx + 5\int x^{-2}\,dx$$

$$= 2e^x + 5\left(-x^{-1}\right) + C$$

$$= 2e^x - \frac{5}{x} + C$$

4. A

We can use the substitution method here because the integrand is made up of two functions related by differentiation. Let:

$$u = x^3 + 1$$
$$du = 3x^2 dx$$
$$\frac{1}{3} du = x^2 dx$$

Therefore,

$$\int x^2 \left(x^3 + 1 \right)^5 dx = \int \left(\underbrace{x^3 + 1}_{u} \right)^5 \underbrace{x^2 dx}_{\frac{1}{3} du}$$

$$= \frac{1}{3} \int u^5 \ du$$

$$= \frac{1}{3} \cdot \frac{u^6}{6} + C$$

$$= \frac{1}{18} \cdot u^6 + C$$

$$= \frac{1}{18} \left(x^3 + 1 \right)^6 + C$$

5. D

We use integration by parts to evaluate the integral. Let:

$$u = x \qquad \text{and} \qquad dv = \sec^2 x \ dx$$
$$du = dx \qquad\qquad\qquad v = \tan x$$

Therefore,

$$\int x \sec^2 x \ dx = x \tan x - \int \tan x \ dx$$
$$= x \tan x - \ln|\sec x| + C$$

6. A

We are given that F is an antiderivative of f. Thus, by the Fundamental Theorem of Calculus, $\int_1^3 f(x) dx = F(3) - F(1)$. Filling in the values for $F(3)$ and $F(1)$ from the chart, we obtain $\int_1^3 f(x) dx = F(3) - F(1) = 8 - 3 = 5$.

7. A

To apply the FTC (Fundamental Theorem of Calculus), we first need to find an antiderivative of $\sec^2\theta$. It's possible to proceed directly because the antiderivative of $\sec^2\theta$ is $\tan\theta$, because $\dfrac{d}{d\theta}(\tan\theta) = \sec^2\theta$. We apply the FTC to get

$$\int_{\frac{5\pi}{6}}^{\frac{5\pi}{4}} \sec^2\theta\ d\theta = \tan\theta \Big|_{\frac{5\pi}{6}}^{\frac{5\pi}{4}} = \tan\left(\frac{5\pi}{4}\right) - \tan\left(\frac{5\pi}{6}\right) = 1 - \left(-\frac{1}{\sqrt{3}}\right) = 1 + \frac{\sqrt{3}}{3}.$$

8. D

To apply the FTC, we first need to find an antiderivative of $\left(3x - \sqrt{2}\right)^5$. Although it is possible to compute this antiderivative directly, in this case it's sensible to make a u-substitution. Let:

$$u = \left(3x - \sqrt{2}\right)$$
$$du = 3\ dx$$
$$\frac{1}{3}du = dx$$

Using the properties of definite integrals, we can rewrite the intergral as:

$$\int_{0}^{\sqrt{2}} \left(3x - \sqrt{2}\right)^5 dx = \int_{x=0}^{x=\sqrt{2}} \left(\underbrace{3x - \sqrt{2}}_{u}\right)^5 \underbrace{dx}_{\frac{1}{3}du}$$

At this stage, our limits of integration are given as x-values. We can transform them to u-values, changing variables completely and eliminating the need to substitute back at the end of the problem.

Because $u = \left(3x - \sqrt{2}\right)$, when $x = \sqrt{2}$, $u = 3\sqrt{2} - \sqrt{2} = 2\sqrt{2}$ and when $x = 0$, $u = -\sqrt{2}$.

We rewrite the integral—changing the variable, including the limits of integration:

$$\int\limits_{0}^{\sqrt{2}} \left(3x - \sqrt{2}\right)^5 dx = \int\limits_{x=0}^{x=\sqrt{2}} \left(\underbrace{3x - \sqrt{2}}_{u}\right)^5 \underbrace{\frac{dx}{\frac{1}{3}\,du}}$$

$$= \frac{1}{3} \int\limits_{-\sqrt{2}}^{2\sqrt{2}} \left(u\right)^5 du$$

The antiderivative of u^5 is $\dfrac{u^6}{6}$ because $\dfrac{d}{du}\left(\dfrac{u^6}{6}\right) = u^5$.

We apply the FTC:

$$\frac{1}{3} \int\limits_{-\sqrt{2}}^{2\sqrt{2}} \left(u\right)^5 du = \frac{1}{3}\left[\frac{u^6}{6}\right]_{-\sqrt{2}}^{2\sqrt{2}}$$

$$= \frac{1}{3}\left[\frac{\left(2\sqrt{2}\right)^6}{6} - \frac{\left(-\sqrt{2}\right)^6}{6}\right]$$

$$= \frac{1}{3}\left(\frac{512}{6} - \frac{8}{6}\right)$$

$$= 28$$

9. C

To apply the FTC, we first need to find an antiderivative of the integrand. Right now, it looks hopelessly complicated, but if we divide the denominator into the numerator, we get an expression that we can find an antiderivative for:

$$\frac{8x^3 - 12x^2 + 7x + 2}{2x - 2} = 4x^2 - 2x + \frac{3}{2} + \frac{5}{2x - 2}$$

We rewrite the initial integral as:

$$\int\limits_{-1}^{0} \frac{8x^3 - 12x^2 + 7x + 2}{2x - 2}\, dx = \int\limits_{-1}^{0} 4x^2 - 2x + \frac{3}{2} + \frac{5}{2x - 2}\, dx$$

Using properties of the definite integral, we write

$$\int\limits_{-1}^{0} 4x^2 - 2x + \frac{3}{2} + \frac{5}{2x - 2}\, dx = 4\int\limits_{-1}^{0} x^2\, dx - 2\int\limits_{-1}^{0} x\, dx + \int\limits_{-1}^{0} \frac{3}{2}\, dx + \frac{5}{2}\int\limits_{-1}^{0} \frac{1}{x - 1}\, dx$$

To find the antiderivatives in the first three integrals, we use the fact that the antiderivative of x^r is $\dfrac{x^{r+1}}{r+1}$. To find the antiderivative in the fourth integral, we can use u-substitution ($u = x - 1$) or we can compute directly that an antiderivative of $\dfrac{1}{x-1}$ is $\ln|x-1|$.

We now apply the FTC:

$$\int_{-1}^{0} 4x^2 - 2x + \frac{3}{2} + \frac{5}{2x-2}\,dx = 4\int_{-1}^{0} x^2\,dx - 2\int_{-1}^{0} x\,dx + \int_{-1}^{0} \frac{3}{2}\,dx + \frac{5}{2}\int_{-1}^{0} \frac{1}{x-1}\,dx$$

$$= 4\left[\frac{x^3}{3}\right]_{-1}^{0} - 2\left[\frac{x^2}{2}\right]_{-1}^{0} + \frac{3}{2}[x]_{-1}^{0} + \frac{5}{2}\Big[\ln|x-1|\Big]_{-1}^{0}$$

$$= 4\left(0 - \frac{(-1)^3}{3}\right) - 2\left(0 - \frac{(-1)^2}{2}\right) + \frac{3}{2}(0 - (-1)) + \frac{5}{2}\big(\ln|0-1| - \ln|(-1)-1|\big)$$

$$= \frac{23}{6} - \frac{5}{2}\ln 2$$

10. E

Use the integration by parts method and let:

$$u = \ln x \qquad dv = x^2\,dx$$
$$du = \frac{1}{x}\,dx \quad \text{and} \quad v = \frac{1}{3}x^3$$

We rewrite the integral

$$\int x^2 \ln x\,dx = \frac{1}{3}x^3 \ln x - \int \left(\frac{1}{3}x^3\right)\left(\frac{1}{x}\right)dx$$

$$= \frac{1}{3}x^3 \ln x - \frac{1}{3}\int x^2\,dx$$

$$= \frac{1}{3}x^3 \ln x - \frac{1}{3}\cdot\frac{1}{3}x^3 + C$$

$$= \frac{1}{3}x^3 \ln x - \frac{1}{9}x^3 + C$$

Therefore,

$$\int_1^e x^2 \ln x \, dx = \left[\frac{1}{3} x^3 \ln x - \frac{1}{9} x^3 \right]_1^e$$

$$= \left[\frac{1}{3}(e)^3 \ln e - \frac{1}{9}(e)^3 \right] - \left[\frac{1}{3}(1)^3 \ln 1 - \frac{1}{9}(1)^3 \right]$$

$$= \left[\frac{1}{3}(e)^3 - \frac{1}{9}(e)^3 \right] - \left[0 - \frac{1}{9} \right]$$

$$= \frac{1}{9}\left(2e^3 + 1 \right)$$

Applications of the Definite Integral

WHAT ARE APPLICATIONS OF THE DEFINITE INTEGRAL?

In the last chapter, we learned how to compute definite integrals using the Fundamental Theorem of Calculus. In this chapter, we will apply what we learned to solve problems involving area and volume.

CONCEPTS TO HELP YOU

1. Area under a curve: The definite integral is the area under a curve and we can compute this area on the interval $a \leq x \leq b$ using the Fundamental Theorem of Calculus. We only need to find a function F whose derivative is f and compute $F(b) - F(a)$:

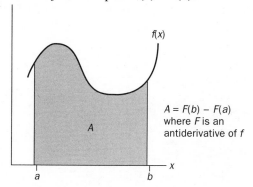

$A = F(b) - F(a)$
where F is an
antiderivative of f

2. Area between two curves: The area between two curves is the integral of the length of the typical cross section:

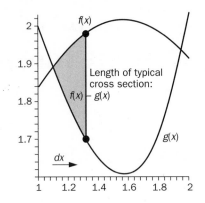

3. Volume of solids: The volume of a solid is the integral of the area of the typical cross section:

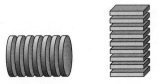

4. Solids of revolution: A solid of revolution is created when a region is rotated around a line, usually the *x*-axis, the *y*-axis, or a line parallel to one of these axes.

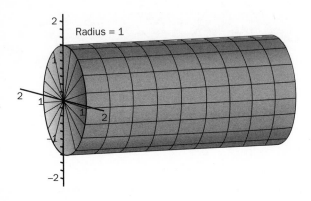

STEPS YOU NEED TO REMEMBER

1. Find the typical cross section.

For area problems, the typical cross section is a line from the curve to the x-axis or from one curve to the other curve.

For volume problems, the typical cross section is a polygon or a circle.

2. Find the limits of integration.

If an interval is not given, the limits of integration are the points where the curves intersect.

3. Integrate the typical cross section.

$$\text{Area} = \int_a^b \left(\text{Length of the typical cross section}\right) dx$$

$$\text{Volume} = \int_a^b \left(\text{Area of the typical cross section}\right) dx$$

COMMON APPLICATIONS OF DEFINITE INTEGRALS QUESTIONS

Area between a Curve and the x-Axis: Find the area between the curve $f(x) = x^3 + x^2 - 6x$ and the x-axis on the interval $[-3,2]$.

Step 1: Graph the function.

Step 2: Shade the area between the function and the x-axis on the given interval $[a,b]$.

Step 3: Evaluate the definite integral $\int_a^b f(x)\, dx$ using the Fundamental Theorem of Calculus.

Solution and Explanation: The typical cross section of the area between $f(x)$ and the x-axis is a line from the x-axis to $f(x)$. Therefore, the area between $f(x)$ and the x-axis is given by the definite integral $\int_{-3}^{2} \left(x^3 + x^2 - 6x\right) dx$. When we graph $f(x) = x^3 + x^2 - 6x$, we see that the graph is above the x-axis on $[-3,0]$ and the graph is below the x-axis on $[0,2]$.

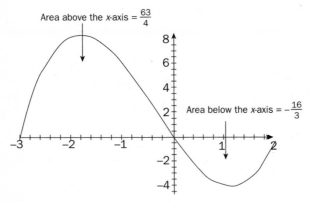

Area above the x-axis = $\frac{63}{4}$

Area below the x-axis = $-\frac{16}{3}$

The area above the x-axis will be positive and the area below the x-axis will be negative.

$$\int_{-3}^{2}\left(x^3+x^2-6x\right)dx = \underbrace{\int_{-3}^{0}\left(x^3+x^2-6x\right)dx}_{\text{area above the }x\text{-axis}} + \underbrace{\int_{0}^{2}\left(x^3+x^2-6x\right)dx}_{\text{area below the }x\text{-axis}}$$

$$=\left[\frac{1}{4}x^4+\frac{1}{3}x^3-3x^2\right]_{-3}^{0}+\left[\frac{1}{4}x^4+\frac{1}{3}x^3-3x^2\right]_{0}^{2}$$

$$=\frac{63}{4}+\left(-\frac{16}{3}\right)$$

$$=\frac{125}{12}$$

We can also use the definite integral to evaluate the area between two curves.

Area between Two Curves: Find the area between the curves $f(x) = \sin x$ and $g(x) = \cos x$ on the interval $[0,2\pi]$.

Step 1: Graph the functions on the same set of axes.

Step 2: Shade the area within the interval.

Step 3: Evaluate the integral $\int_{a}^{b}$ [(function on top) minus (the function on the bottom)] dx.

Solution and Explanation: The typical cross section here would be a line between the two functions. To find the area between the curves $f(x) = \sin x$ and $g(x) = \cos x$ on the interval $[0,2\pi]$, we need to determine which curve is on top and which is on the bottom to compute $\int_a^b$ [(function on top) minus (the function on the bottom)] dx.

When we graph the two functions, we see that $g(x)$ starts out on top, but at some point it moves to the bottom and $f(x)$ moves to the top. Then, further along on the x-axis, $g(x)$ moves back to the top. When this happens, we must break the integral into pieces.

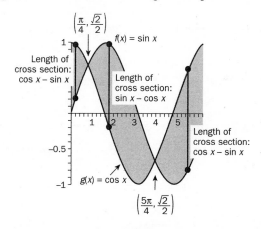

The shaded area is the area we are trying to find.

We know that on the interval $[0,2\pi]$, $\sin x$ and $\cos x$ intersect at $x = \frac{\pi}{4}$ and $x = \frac{5\pi}{4}$. We see from looking at the graph that, on the interval $\left[0, \frac{\pi}{4}\right]$, the cosine curve lies on top; on $\left[\frac{\pi}{4}, \frac{5\pi}{4}\right]$, the sine curve is on top and on $\left[\frac{5\pi}{4}, 2\pi\right]$, $\cos x$ is back on top. The area between the two curves is therefore expressed by the sum of three definite integrals:

$$\int_0^{\frac{\pi}{4}} \cos x - \sin x \ dx + \int_{\frac{\pi}{4}}^{\frac{5\pi}{4}} \sin x - \cos x \ dx + \int_{\frac{5\pi}{4}}^{2\pi} \cos x - \sin x \ dx$$

$$\int_0^{\frac{\pi}{4}} \cos x - \sin x \ dx = \sqrt{2} - 1$$

$$\int\limits_{\frac{\pi}{4}}^{\frac{5\pi}{4}} \sin x - \cos x \ dx = 2\sqrt{2}$$

$$\int\limits_{\frac{5\pi}{4}}^{2\pi} \cos x - \sin x \ dx = 1 + \sqrt{2}$$

The sum of these three definite integrals is $4\sqrt{2}$.

Another type of problem involving area is the area of a region bounded by two or more functions and the *x*-axis.

Area Bounded by Curves: The region *R* shown below is bounded by the curves *f*(*x*) = *x*, *g*(*x*) = −(*x* − 1)² + 1, and the *x*-axis. Find the area of the region *R*.

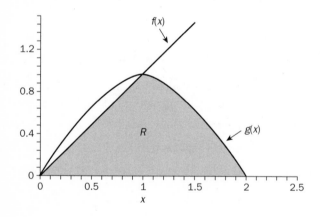

Step 1: Graph the function on the same set of axes.

Step 2: Determine the limits of integration.

Step 3: Evaluate the definite integrals.

Solution and Explanation: To find the area bound by curves, we need to examine the graph of the curves to determine what the limits of integration are.

Looking at the graph, we see that there are two distinct regions. The left region R_1 starts at the point $x = 0$ and runs until the two curves intersect at $x = 1$. The second region R_2 runs from $x = 1$ to the point where $g(x)$ intersects the x-axis at $x = 2$.

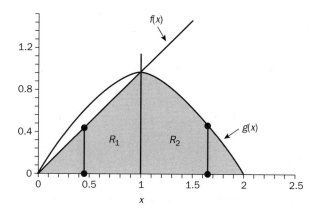

The region R_1 lies between the curve $f(x) = x$ and the x-axis. Therefore, the area of region R_1 is computed by:

$$\text{Area of } R_1 = \int_0^1 x \ dx = \frac{1}{2}x^2 \Big|_0^1 = \frac{1}{2}(1)^2 - \frac{1}{2}(0)^2 = \frac{1}{2}$$

The region R_2 lies between the curve $g(x) = -(x-1)^2 + 1$ and the x-axis. Therefore, the area of region R_2 is computed by:

$$\text{Area of } R_2 = \int_1^2 -(x-1)^2 + 1 \ dx$$

$$= \int_1^2 -\left(x^2 - 2x + 1\right) + 1 \ dx$$

$$= \int_1^2 -x^2 + 2x - 1 + 1 \ dx$$

$$= \int_1^2 -x^2 + 2x \ dx$$

$$= -\frac{x^3}{3} + x^2 \ \Big|_1^2$$

$$= \left(-\frac{8}{3} + 2\right) - \left(-\frac{1}{3} + 1\right)$$

$$= \frac{2}{3}$$

The area of R is equal to the sum of the areas of R_1 and R_2:

$$\text{Area of } R = \text{Area of } R_1 + \text{Area of } R_2$$

$$= \int_0^1 x \ dx + \int_1^2 -(x-1)^2 + 1 \ dx$$

$$= \frac{1}{2} + \frac{2}{3}$$

$$= \frac{7}{6}$$

We can think of the area between two curves as being swept out by the typical cross section, and we can think of volumes the same way. We can think of a volume of a solid as being swept out by the typical cross section, i.e., an area.

Volume of Solids: Suppose the region inside the circle $x^2 + y^2 = 9$ is the base of a solid.

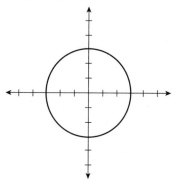

If each cross section of the solid perpendicular to the x-axis is a square, what is the volume of the solid?

Step 1: Draw the typical cross section.

Step 2: If the cross section is perpendicular to the x-axis, express the area of the cross section in terms of x. If the cross section is perpendicular to the y-axis, express the area of the cross section in terms of y.

Step 3: Evaluate the volume of the solid as $V = \int_a^b A(x)\,dx$ or $V = \int_a^b A(y)\,dy$.

> **Solution and Explanation:** The figures below illustrate what a solid with a circular base and square cross sections look like:

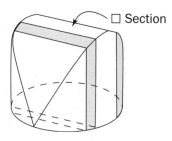 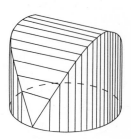

□ Section

We are told that the cross sections are perpendicular to the x-axis. Start by drawing the typical cross section:

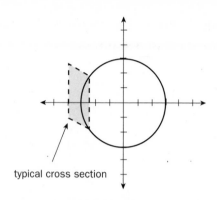

typical cross section

As you can see, the length of the side of a square cross section is the distance from a point on the circle above the *x*-axis to its opposite point below the *x*-axis. It follows from $x^2 + y^2 = 9$ that $y = \sqrt{9 - x^2}$. This means that the distance from the *x*-axis to a point on the circle is $\sqrt{9 - x^2}$, so the length of the side of a square cross section is $s = 2\sqrt{9 - x^2}$.

The area of an arbitrary square cross section is thus

$$A(x) = s^2 = \left(2\sqrt{9 - x^2}\right)^2 = 4\left(9 - x^2\right)$$

and the integral of the area is

$$V = \int_a^b A(x)\,dx$$

$$= \int_{-3}^3 4\left(9 - x^2\right)dx$$

$$= 4\left[9x - \frac{1}{3}x^3\right]_{-3}^3$$

$$= 4\left[18 - (-18)\right]$$

$$= 144$$

The volume of the solid is 144.

We can also use the definite integral to find the volume of a solid of revolution.

Volume of Solids of Revolution: Consider the region bounded by $y = x^2$, $x = 1$, $x = 2$, and the *x*-axis.

What is the volume of the solid that results from revolving this region around
(a) *x*-axis? (b) line $x = 4$?

Step 1: Draw a picture of the region *R* that is being rotated.

Step 2: To the best of your ability, draw a picture of the solid of revolution.

Step 3: Draw the typical cross section.

Step 4: Determine its shape (circle or washer).

Step 5: Determine the radius of the circle (or, for a washer, the radii of the outer and inner circles).

Step 6: Determine the formula for $A(x)$, the area of the typical cross section in terms of *x*.

Step 7: Determine the boundaries of the region *R*, $x = a$, and $x = b$.

Step 8: Compute the definite integral of $A(x)$ with limits *a* and *b*.

> **Solution and Explanation:** The cross sections of solids of revolution are always circles or washers (i.e., a circle with a hole in the middle). The formula for the area of a circle is $A = \pi \cdot r^2$.
>
> (a) We start by graphing the functions:

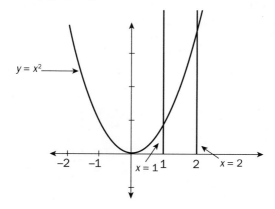

> The region bounded by the curve and the lines is the portion we will rotate around the *x*-axis:

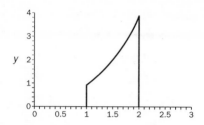

We can draw a two-dimensional slice of the solid of revolution to get a sense of what the solid of revolution looks like:

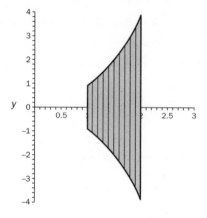

Trying to draw a picture of the three-dimensional solid can be tricky. It isn't necessary to draw the three-dimensional solid, but it can't hurt.

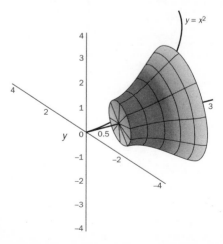

The radius of the typical cross section is x^2 so the area of the cross section is $A(x) = \pi(x^2)^2$:

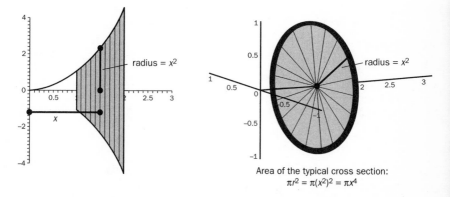

Area of the typical cross section:
$$\pi r^2 = \pi(x^2)^2 = \pi x^4$$

The boundaries of the solid are $x = 1$ and $x = 2$, so the volume of the solid is:

$$\int_1^2 \pi x^4 \ dx = \frac{31\pi}{5}$$

(b) A solid of revolution is obtained when a curve is revolved around *any* line, not just the x-axis. When we revolve the region around the line $x = 4$, the solid we obtain has a hole in the middle.

The typical cross section here is not a circle; it is a washer, or, if we prefer, a donut. The area of a washer is the area of the outer circle minus the area of the inner circle. Let's call the radius of the outer circle R and the radius of the inner circle r. The area of the washer is thus defined as follows:

Area of washer = Area of outside circle − Area of inside circle
$= \pi R^2 - \pi r^2$, where R is the radius of the outer circle and r is the radius of the inner circle.

In this problem, the radius of the outside circle is the distance from curve $y = x^2$ to the axis of rotation, $x = -1$, which is $R = x^2 - (-1) = x^2 + 1$.

The radius of the inner circle is the distance from the x-axis to the axis of rotation $x = -1$:

$r = 0 - (-1) = 1.$

The area of the typical cross section, which is a washer, is:

$$A(x) = \pi R^2 - \pi r^2$$

$$= \pi \left(x^2 + 1 \right)^2 - \pi \cdot 1^2$$

$$= \pi \left[\left(x^2 + 1 \right)^2 - 1 \right]$$

The volume of the solid is the integral of the area of the typical cross section. Because the boundaries of the region are $x = 1$ and $x = 2$, the volume of the solid is:

$$\int_1^2 \pi \left[\left(x^2 + 1 \right)^2 - 1 \right] dx = \frac{163\pi}{15}$$

CHAPTER QUIZ

1. Using the definition of the definite integral as the area under the curve, compute $\int_0^6 (2x - 2) dx$.

   (A)   36

   (B)   18

   (C) −18

   (D)   24

   (E) −24

2. Find the area between the curves $f(x) = x^2 - 3x + 2 = (x - 1)(x - 2)$ and $g(x) = -(x^2 - 3x + 2) = -(x - 1)(x - 2)$ on the interval from $x = 0$ to $x = 2$.

   (A)   0

   (B) $\frac{2}{3}$

   (C)   1

   (D) $\frac{4}{3}$

   (E)   2

3. Find the area between the curves $f(x) = x^2$ and $g(x) = 8 - x^2$ on the interval $[-2,2]$.

   (A) $\dfrac{46}{3}$

   (B) $\dfrac{64}{3}$

   (C) $\dfrac{1}{3}$

   (D) $\dfrac{50}{3}$

   (E) $\dfrac{5}{3}$

4. Find the area between the curves $f(x) = x^2$ and $g(x) = x^3$ on the interval $[0,1]$.

   (A) 1

   (B) 2

   (C) $\dfrac{1}{12}$

   (D) $\dfrac{1}{2}$

   (E) $\dfrac{1}{16}$

5. Find the area bounded by $f(x) = \sqrt{x}$, $g(x) = x - 2$, and the y-axis.

   (A) $\dfrac{4\sqrt{2}}{3}$

   (B) $\dfrac{1}{2}$

   (C) $2\sqrt{2}$

   (D) $\dfrac{4}{3}$

   (E) $\dfrac{10}{3}$

6. Find the volume of the solid of revolution determined by rotating the area bounded by the graphs of $f(x) = x^2$ and $g(x) = 3x$ about the x-axis.

   (A) $\dfrac{162\pi}{5}$

   (B) $\dfrac{27\pi}{2}$

   (C) 9π

   (D) 28π

   (E) $\dfrac{240\pi}{7}$

7. Consider the region enclosed by the x-axis, the line $x = 1$, the line $x = 3$, and the curve $f(x) = x^{\frac{3}{2}}$. If this region is revolved around the x-axis, the volume of the solid is

   (A) 20π.

   (B) $\dfrac{25}{2}\pi$.

   (C) 21π.

   (D) 4π.

   (E) $\dfrac{100\pi}{3}$.

For Questions 8–10, let R be the region enclosed by the graphs of $f(x) = ax(2-x)$ and $g(x) = ax$ for some positive real number a.

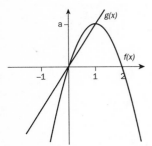

8. Find the area of region R.
   (A) $12a$
   (B) $\dfrac{1}{a}$
   (C) $\dfrac{a}{6}$
   (D) $\dfrac{ax}{6}$
   (E) 6

9. Find the volume of the solid of revolution generated when R is rotated about the x-axis.
   (A) $\dfrac{\pi a^2}{2}$
   (B) $\dfrac{\pi a^2}{3}$
   (C) $\dfrac{\pi a^2}{4}$
   (D) $\dfrac{\pi a^2}{5}$
   (E) $\dfrac{\pi a^2}{6}$

10. Assume a solid exists with a cross-section area of R and uniform thickness π. Find the value of a for which this solid has the same volume as the solid in Question 9.
    (A) $\dfrac{1}{6}$
    (B) $\dfrac{2}{6}$
    (C) $\dfrac{1}{2}$
    (D) $\dfrac{2}{3}$
    (E) $\dfrac{5}{6}$

Answers and Explanations

1. D

The definite integral is the area under the curve, where the area under the x-axis is negative. If we graph this function we can see that there are two regions of area: a small triangle below the x-axis and a large triangle above the x-axis.

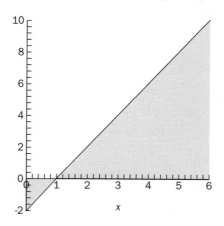

Therefore,

$$\int_0^6 (2x-2)\,dx = \underbrace{\int_0^1 (2x-2)\,dx}_{\text{area of smaller triangle}} + \underbrace{\int_1^6 (2x-2)\,dx}_{\text{area of larger triangle}}$$

$$= \left[x^2 - 2x \right]_0^1 + \left[x^2 - 2x \right]_1^6$$

$$= -1 + 25$$

$$= 24$$

2. E

Start by sketching the graphs of these curves on the same set of axes. Each curve is a parabola that intersects the x-axis at $x = 1$ and $x = 2$. The parabola $f(x)$ opens up and parabola $g(x)$ opens down.

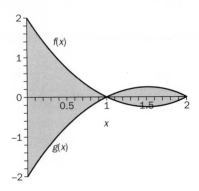

The area between the curves is the integral of the typical cross section, i.e., the top curve minus the bottom curve. From $x = 0$ to $x = 1$, the curve $f(x)$ is on top; from $x = 1$ to $x = 2$, $g(x)$ is on top. The area is therefore given by two integrals:

$$A(x) = \int_{0}^{1} f(x) - g(x) \; dx + \int_{1}^{2} g(x) - f(x) \; dx$$

Notice that $g(x) = -f(x)$. Thus we can simplify the calculation:

$$A(x) = \int_{0}^{1} f(x) - \left(-f(x)\right) \; dx + \int_{1}^{2} \left(-f(x)\right) - f(x) \; dx$$

$$= \int_{0}^{1} 2f(x) \; dx + \int_{1}^{2} -2f(x) \; dx$$

$$= 2\int_{0}^{1} f(x) \; dx - 2\int_{1}^{2} f(x) \; dx$$

$$= 2\int_{0}^{1} x^2 - 3x + 2 \; dx - 2\int_{1}^{2} x^2 - 3x + 2 \; dx$$

$$= 2\left(\frac{x^3}{3} - \frac{3x^2}{2} + 2x \right)\Bigg|_{0}^{1} - 2\left(\frac{x^3}{3} - \frac{3x^2}{2} + 2x \right)\Bigg|_{1}^{2}$$

$$= 2$$

3. B

When we graph the functions, we see that $g(x) = 8 - x^2$ is the function on top.

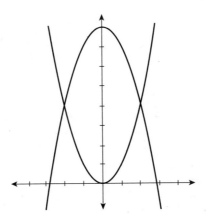

Therefore,

$$A = \int_a^b [(\text{function on top}) - (\text{function on bottom})]\, dx$$

$$= \int_{-2}^{2} \left[\left(8 - x^2\right) - \left(x^2\right) \right] dx$$

$$= \int_{-2}^{2} \left(8 - 2x^2\right) dx$$

$$= \left[8x - \frac{2}{3}x^3 \right]_{-2}^{2}$$

$$= \frac{32}{3} - \left(-\frac{32}{3}\right)$$

$$= \frac{64}{3}$$

4. C

The graph of x^2 is above x^3 on the interval $[0,1]$:

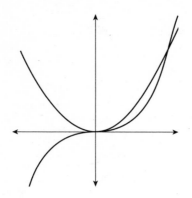

Therefore, we compute the area as

$$\int_0^1 \left(x^2 - x^3\right)dx = \int_0^1 x^2\,dx - \int_0^1 x^3\,dx$$

$$= \frac{1}{3}x^3 \Big|_0^1 \ - \frac{1}{4}x^4 \Big|_0^1$$

$$= \left[\frac{1}{3}x^3 - \frac{1}{4}x^4\right]_0^1$$

$$= \left(\frac{1}{3} - \frac{1}{4}\right) - 0$$

$$= \frac{1}{12}$$

5. D

When we graph $f(x)$ and $g(x)$ on the same axes, we see that we need to divide the area bounded f and g and the y-axis into three smaller areas.

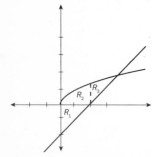

$$R_1 = \int_0^2 (x-2)\,dx$$

$$= \int_0^2 x\,dx - \int_0^2 2\,dx$$

$$= \left[\frac{1}{2}x^2 - 2x \right]_0^2$$

$$= \left[\frac{1}{2}(2)^2 - 2(2) \right] - 0$$

$$= -2$$

$$R_2 = \int_0^2 \sqrt{x}\,dx$$

$$= \int_0^2 x^{\frac{1}{2}}\,dx$$

$$= \frac{2}{3}x^{\frac{3}{2}}\Big|_0^2$$

$$= \frac{2}{3}(2)^{\frac{3}{2}} - 0$$

$$= \frac{4\sqrt{2}}{3}$$

$$R_3 = \int_2^4 \left(\sqrt{x} - (x-2) \right)dx$$

$$= \int_2^4 \sqrt{x}\,dx - \int_2^4 x\,dx + \int_2^4 2\,dx$$

$$= \left[\frac{2}{3}x^{\frac{3}{2}} - \frac{1}{2}x^2 + 2x \right]_2^4$$

$$= \left[\frac{2}{3}(4)^{\frac{3}{2}} - \frac{1}{2}(4)^2 + 2(4) \right] - \left[\frac{2}{3}(2)^{\frac{3}{2}} - \frac{1}{2}(2)^2 + 2(2) \right]$$

$$= \frac{10 - 4\sqrt{2}}{3}$$

The total area is the sum of the three smaller areas:

$$R = R_1 + R_2 + R_3$$

$$= -2 + \frac{4\sqrt{2}}{3} + \frac{10 - 4\sqrt{2}}{3}$$

$$= \frac{4}{3}$$

6. A

First, we see that the graph of $f(x)$ is always below the graph $g(x)$ from $x = 0$ to $x = 3$ (where the two graphs intersect). We then see that the volume in question can be determined using the washer method. The typical washer will have area $\pi((3x)^2 - (x^2)^2)$. Therefore, the volume in question equals

$$\int_0^3 \pi (3x)^2 - \left(x^2\right)^2 dx$$

Note that

$$\int \pi \left((3x)^2 - \left(x^2\right)^2 \right) dx = \pi \int \left(9x^2 - x^4 \right) dx$$

$$= \pi \left[3x^3 - \frac{x^5}{5} \right]$$

Now we evaluate from $x = 0$ to $x = 3$ using the Fundamental Theorem of Calculus and obtain

$$\pi \left(81 - \frac{243}{5} \right) = \frac{162\pi}{5}$$

7. A

The equation to use is $V = \int_a^b \pi [f(x)]^2 \, dx$, where $f(x) = $ radius of cross section given by the function and a and b are the left and right bounds of the surface of revolution. The $\pi [f(x)]^2$ term measures the area of a cross-sectional disk and the $\int_a^b dx$ term adds them up to give the entire volume of the solid. So, $V = \int_1^3 \pi [x^{3/2}]^2 \, dx = \int_1^3 \pi x^3 \, dx = \frac{\pi x^4}{4} \Big|_1^3 = \pi [\frac{3^4}{4} - \frac{1}{4}] = 20 \cdot \pi$

8. C

The functions intersect at $x = 0$ and $x = 1$.

Therefore, the area of the region is given by

$$\int_0^1 f(x) - g(x) dx = \int_0^1 (2ax - ax^2) - ax dx = \int_0^1 ax - ax^2 dx = \frac{ax^2}{2} - \frac{ax^3}{3}\Big|_0^1 = \frac{a}{2} - \frac{a}{3} = \frac{a}{6}$$

9. D

The volume of revolution (about the x-axis) is given by

$$\int_0^1 \pi f^2(x) - \pi g^2(x) dx = \int_0^1 \pi(2ax - ax^2)^2 - \pi(ax)^2 dx = \pi \int_0^1 4a^2 x^2 - 4a^2 x^3 + a^2 x^4 - a^2 x^2 dx$$

$$= \pi \left(a^2 x^3 - a^2 x^4 + \frac{a^2 x^5}{5} \right)\Big|_0^1 = \pi \left(a^2 - a^2 + \frac{a^2}{5} \right) - 0 = \frac{\pi a^2}{5}$$

10. E

The volume created by the cross section R and thickness a is $\left(\frac{a}{6}\right)\pi = \frac{\pi a}{6}$. We need to find the positive value of a for which $\frac{\pi a}{6} = \frac{\pi a^2}{5}$.

Therefore, for $a = \frac{5}{6}$ the volumes are equal.

CHAPTER 11

Max-Min and Other Rates

WHAT ARE MAX-MIN AND OTHER RATES?

Back in Chapter 8, we learned that understanding a function requires us to identify its relative extrema. We will use what we learned about relative extrema to understand the motion of a particle on a straight line. Then we will use our knowledge of derivatives and integrals to solve problems involving position, velocity, and acceleration.

CONCEPTS TO HELP YOU

1. Position function: The position function tells you where an object is along a line at any time t; it is denoted by $s(t)$.

2. Velocity function: Velocity is the rate of change of position and is denoted by $v(t)$.

    - When the particle is at rest, $v(t) = 0$.

    - When the particle is moving right, $v(t) > 0$.

    - When the particle is moving left, $v(t) < 0$.

    - When the particle changes direction, $v(t)$ changes sign.

3. Acceleration function: Acceleration is the rate of change of velocity and is denoted by $a(t)$.

    - When $v(t)$ is increasing, $a(t) > 0$.

    - When $v(t)$ is decreasing, $a(t) < 0$.

    - When $v(t)$ is constant, $a(t) = 0$.

STEPS YOU NEED TO REMEMBER

1. Finding position.

Integrate the velocity function once or integrate the acceleration function twice to arrive at the position function:

$$s(t) = \int v(t)\,dt = \int \left[\int a(t)\,dt \right] dt$$

2. Finding velocity.

Differentiate the position function or integrate the acceleration function to arrive at the velocity function:

$$s'(t) = v(t) = \int a(t)\,dt$$

3. Finding acceleration.

Differentiate the position function twice or differentiate the velocity function once to arrive at the acceleration function:

$$s''(t) = v'(t) = a(t)$$

4. Total distance traveled.

Compute the definite integral of the position function:

$$\int_a^b s(t)\,dt = S(b) - S(a), \text{ where } S(t) \text{ is the antiderivative of } s(t)$$

COMMON MAX-MIN AND OTHER RATES QUESTIONS

For problems 1–3 below, the position of a particle along a horizontal line at $t \geq 0$ is given by $s(t) = t^3 - 6t^2 + 9t + 1$, where t is measured in seconds and s is measured in feet.

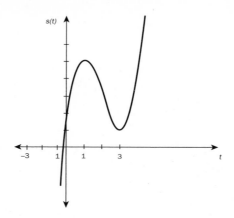

Particle at Rest: Find any time(s) when the particle is at rest.

Step 1: Find the velocity function.

Step 2: Set $v(t) = 0$.

Step 3: Solve for t.

> **Solution and Explanation:** The particle is at rest when $v(t) = 0$. We are given only the position function, so we need to find the velocity function. Velocity is the instantaneous rate of change of position; i.e., $v(t) = s'(t)$.
>
> $$s(t) = t^3 - 6t^2 + 9t + 1$$
> $$s'(t) = 3t^2 - 12t + 9$$
> $$v(t) = 3t^2 - 12t + 9$$
>
> To find the time(s) when the particle is at rest, we set $v(t) = 0$:
>
> $$3t^2 - 12t + 9 = 0$$
> $$(3t - 9)(t - 1) = 0$$
>
> $$3t - 9 = 0 \quad \text{or} \quad t - 1 = 0$$
> $$t = 3 \qquad\qquad t = 1$$
>
> Therefore, at $t = 1$ and $t = 3$, the particle is at rest.

What we have just done is very similar to the work we did in Chapter 8 to find the critical points of a curve.

Recall that the slope of the tangent line to a curve is the derivative of the curve. This slope is 0 (i.e., the line is horizontal) at the critical points of the curve.

Change in Direction: Find any time(s) when the particle changes direction.

Step 1: Set $v(t) = 0$ and solve for t.

Step 2: Do a sign analysis of $v(t)$ to determine when the particle changes direction.

Solution and Explanation: It is only possible for $v(t)$ to change direction when $v(t) = 0$, because the particle is at rest immediately before it turns around and heads in the opposite direction.

We already know that the particle is at rest at $t = 1$ and $t = 3$. To determine whether the particle changes direction at these times, we do a sign analysis just like we did when we performed the First Derivative Test back in Chapter 8.

Construct a chart to analyze the sign of $v(t) = 3t^2 - 12t + 9$ at certain intervals.

Interval	$0 < t < 1$	$1 < t < 3$	$t > 3$
Test Point	0	2	4
Sign of $v(t)$	$v(0) = 9$	$v(2) = -3$	$v(4) = 9$
Conclusion	positive (right)	negative (left)	positive (right)

The chart shows us that the particle starts out moving right (positive) from its initial position. After the first second, it turns around and moves left (negative). Two seconds later, at $t = 3$, it turns around again and heads back in the positive direction.

Even though the motion of the particle is along a horizontal line, we can see that our sign analysis corresponds to the graph of $s(t)$ above, according to which the curve changes direction at $t = 1$ and at $t = 3$.

Constant Velocity: Find the time(s) when the velocity is constant.

Step 1: Find acceleration function $a(t)$.

Step 2: Set $a(t) = 0$.

Step 3: Solve for t.

> **Solution and Explanation:** Acceleration is the rate of change of velocity; i.e., $a(t) = v'(t)$.
>
> $$v(t) = 3t^2 - 12t + 9$$
> $$v'(t) = 6t - 12$$
> $$a(t) = 6t - 12$$

When velocity is increasing with respect to time, acceleration is positive. When velocity is decreasing with respect to time, acceleration is negative. Velocity is constant when acceleration is zero.

The acceleration is 0 at $t = 2$:

$$6t - 12 = 0$$
$$t = 2$$

When we set $a(t) = 0$, we found the point of inflection of $s(t)$. Recall that the inflection point is the point at which the concavity of a function changes, and we determine the concavity of a function by applying the Second Derivative Test. Since $a(t)$ is the *second derivative* of $s(t)$, solving for $a(t) = 0$ gives us the point at which the acceleration of the particle changes. At $t = 2$, the acceleration of the particle changes from decreasing to increasing:

Interval	$0 < t < 2$	$t > 2$
Test Point	1	3
Sign of $a(t)$	$a(1) = -6$	$a(3) = 6$
Conclusion	decreasing	increasing

At $t = 2$, the velocity is

$$v(t) = 3t^2 - 12t + 9$$
$$v(2) = 3(2)^2 - 12(2) + 9$$
$$= -3 \text{ feet/sec}$$

In problems 1–3, we used derivatives to solve motion problems. Now we will see how we can use integrals to solve these types of problems.

In problems 4 and 5, a particle moves along the horizontal line with initial position $s(0) = -2$. The velocity of the particle at time $t \geq 0$ is given by $v(t) = t^2 - 3t + 2$.

Position of the Particle: What is the position of the particle at time $t = 3$?

Step 1: Compute the indefinite integral of $v(t)$.

Step 2: Solve for C to get the position function $s(t)$.

Step 3: Find the position by plugging the value of t into $s(t)$.

Solution and Explanation: Because velocity is the derivative of position, the position of the particle at time t is given by:

$$s(t) = \int v(t)\,dt = \int \left(t^2 - 3t + 2\right) dt = \frac{1}{3}t^3 - \frac{3}{2}t^2 + 2t + C$$

To determine $s(t)$, we need to find C. We are told that the initial position of the particle is $s(0) = -2$. We solve

$$\frac{1}{3}(0)^3 - \frac{3}{2}(0)^2 + 2(0) + C = -2$$

$$C = -2$$

Therefore, the position of the particle at any time t is given by

$$s(t) = \frac{1}{3}t^3 - \frac{3}{2}t^2 + 2t - 2.$$

The position of the particle at time $t = 3$ is

$$s(3) = \frac{1}{3}(3)^3 - \frac{3}{2}(3)^2 + 2(3) - 2 = -\frac{1}{2}.$$

Total Distance Traveled: What is the total distance traveled by the particle over the time period $0 \leq t \leq 3$?

Step 1: Find $s(t)$.

Step 2: Calculate the definite integral of $s(t)$ on the given interval.

> **Solution and Explanation:** From the previous problem, we know that the position function is $x(t) = \frac{1}{3}t^3 - \frac{3}{2}t^2 + 2t - 2$.
>
> The total distance traveled in the first three seconds is:
>
> $$\int_0^3 t^2 - 3t + 2 \, dt = \left[\frac{1}{3}t^3 - \frac{3}{2}t^2 + 2t \right]_0^3$$
> $$= \left(\frac{1}{3} \cdot 3^3 - \frac{3}{2} \cdot 3^2 + 2 \cdot 3 \right) - \left(\frac{1}{3} \cdot 0^3 - \frac{3}{2} \cdot 0^2 + 2 \cdot 0 \right)$$
> $$= \frac{3}{2}$$

CHAPTER QUIZ

For problems 1–3, the height of a particle above the ground at time *t* is given by the graph below.

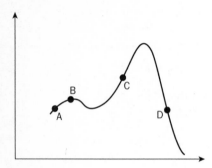

1. At which of the labeled points on the graph is the particle's instantaneous velocity the greatest?

   (A) A

   (B) B

   (C) C

   (D) D

   (E) Cannot be determined

2. At which of the labeled points on the graph is the particle changing directions?

   (A) A

   (B) B

   (C) C

   (D) D

   (E) Cannot be determined

3. At which of the labeled points on the graph is the particle's instantaneous velocity zero?

   (A) A

   (B) B

   (C) C

   (D) D

   (E) Cannot be determined

For problems 4 and 5, a particle moves along the y-axis with velocity
$v(t) = -\dfrac{2}{\pi}\sin\left(\dfrac{\pi}{2}t\right)$ cm/sec for time $t \geq 0$ in seconds.

4. What is its velocity at $t = \dfrac{1}{3}$?

   (A)   1

   (B) π

   (C) $-\dfrac{1}{\pi}$

   (D) $\dfrac{1}{\pi}$

   (E) $-\pi$

5. Find the earliest time, $t_1 > 0$, when the particle changes direction.

   (A) 1

   (B) 2

   (C) 3

   (D) 4

   (E) 5

6. A particle moves along the x-axis so that its velocity at time t is given by the function shown in the graph below. At which time is the particle farthest from the origin?

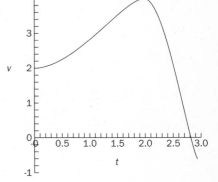

   (A) $t = 0.5$

   (B) $t = 1.5$

   (C) $t = 2$

   (D) $t = 2.8$

   (E) $t = 3$

7. A particle moves along the x-axis with a velocity given by $v(t) = 2 + \sin t$. When $t = 0$, the particle is at $x = -2$. Where is the particle when $t = \pi$?

   (A) π

   (B) 2π

   (C) $\pi - 1$

   (D) $\pi - 2$

   (E) $\pi + 1$

For problems 8 and 9, a particle moves along the x-axis so that its velocity at any time $t > 0$ is given by $v(t) = 5t^2 - 4t + 7$. The position of the particle, $s(t)$ is 8 for $t = 3$.

8. What is the position function of the particle at any time $t \geq 0$?

    (A) $s(t) = \frac{5}{3}t^3 - 2t^2 + 7t - 10$

    (B) $s(t) = \frac{5}{3}t^3 - 2t^2 + 9t - 40$

    (C) $s(t) = \frac{5}{3}t^3 - \frac{2}{3}t^2 + 7t - 40$

    (D) $s(t) = 5t^3 - 2t^2 + 7t - 40$

    (E) $s(t) = \frac{5}{3}t^3 - 2t^2 + 7t - 40$

9. Find the total distance traveled by the particle from time $t = 0$ until the time $t = 2$.

    (A) $\frac{1}{3}$

    (B) $\frac{19}{3}$

    (C) $\frac{58}{3}$

    (D) 20

    (E) 35

10. A particle moves along the x-axis according to the position function

    $$s(t) = \frac{1}{2}t^3 - 5t^2 + 3t + 6,$$

    where t is measured in seconds. At what time t is the acceleration of the particle equal to -1 ft/sec^2?

    (A) 1

    (B) 2

    (C) 3

    (D) 4

    (E) 5

Answers and Explanations

1. C

The particle's instantaneous velocity is greatest at the point where the slope of the tangent line is greatest. We draw the line tangent to the curve at each of the labeled points. By inspection, the slope of the line tangent to the curve at point C is the greatest. Therefore, the particle's instantaneous velocity is greatest at point C.

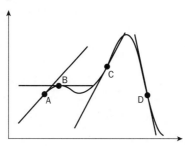

2. B

The particle changes direction at point B. Before this point the height of the particle is increasing, so the particle is moving away from the ground. After this point, the height of the particle is decreasing, so the particle is moving toward the ground. Therefore, the particle changes direction at point B.

3. B

The instantaneous velocity of the particle is given by the slope of the tangent line. The slope of the line tangent to the curve at point B is zero, and so the instantaneous velocity of the particle at this point is zero.

4. C

We can determine the particle's direction by checking the sign of v. If v is positive, the position, or the height of the particle, is increasing, so it is moving up. If v is negative, the height of the particle is decreasing, so it is moving down. We compute:

$$v\left(\frac{1}{3}\right) = -\frac{2}{\pi}\sin\left(\frac{\pi}{2}\cdot\frac{1}{3}\right) = -\frac{2}{\pi}\sin\left(\frac{\pi}{6}\right) = -\frac{2}{\pi}\cdot\frac{1}{2} = -\frac{1}{\pi}$$

The velocity is $-\dfrac{1}{\pi}$ cm/sec. Because it is negative, the particle is moving down.

5. B

The particle changes direction when the velocity changes. Because the velocity function is defined everywhere, this can only happen when $v(t) = 0$. We set $v(t) = 0$ and solve this equation:

$$-\frac{2}{\pi}\sin\left(\frac{\pi}{2}t\right) = 0$$

$$\sin\left(\frac{\pi}{2}t\right) = 0$$

$$\frac{\pi}{2}t = n\pi$$

$$t = 2n, \text{ where } n \text{ is an integer}$$

According to the question, we need $t > 0$, which means that the earliest possibility is when $n = 1$, i.e., $t = 2$ seconds. We now need to confirm that the velocity actually changes sign at $t = 2$, i.e., that this is a relative extremum and not merely a critical point.

To do this, we can do a sign analysis of $v(t)$ or apply the Second Derivative Test to $t = 2$. To apply the Second Derivative Test, we compute $v'(t)$ and evaluate $v'(2)$:

$$v(t) = -\frac{2}{\pi}\sin\left(\frac{\pi}{2}t\right)$$

$$v'(t) = -\frac{2}{\pi} \cdot \underbrace{\frac{\pi}{2}\cos\frac{\pi}{2}t}_{\text{chain rule}}$$

$$= -\cos\frac{\pi}{2}t$$

$$v'(2) = -\cos\pi$$

$$= -1$$

Since the second derivative is negative, a relative maximum occurs at $t = 2$ and the point changes direction at $t = 2$.

6. D

Velocity is the rate of change of distance, so the area under the velocity curve from $t = 0$ to a time t on the graph is the change in distance over the time interval (counting area under the axis as negative):

$$s(t) = \int_0^t v(t) dt$$

Area under the *t*-axis represents distance traveled *away* from the origin. Area above the *t*-axis represents distance traveled *toward* the origin. The distance from the origin is greatest when the definite integral is the greatest; i.e., when the total area (counting area beneath the *x*-axis as negative) is greatest. This occurs at the point $t = 2.8$.

7. B

The position of the particle $s(t)$ is an antiderivative of the velocity function, so we start by computing the indefinit integral of the velocity:

$$v(t) = 2 + \sin t$$
$$s(t) = \int (2 + \sin t) dt$$
$$= 2t - \cos t + C$$

We can use the initial condition to determine C:

$$s(0) = -2$$
$$2(0) - \cos(0) + C = -2$$
$$0 - 1 + C = -2$$
$$C = -1$$

Our position equation is $s(t) = 2t - \cos t - 1$.

Now we evaluate the position function at time $t = \pi$:

$$s(\pi) = 2(\pi) - \cos \pi - 1$$
$$= 2\pi - (-1) - 1$$
$$= 2\pi$$

8. E

The position of the particle is given by

$$s(t) = \int v(t) dt = \int 5t^2 - 4t + 7 dt = \frac{5}{3} t^3 - 2t^2 + 7t + C.$$

Solve for C by substituting $s(3) = 8$:

$$8 = \frac{5}{3}(27) - 2(9) + 21 + C$$
$$= 45 - 18 + 21 + C$$
$$= 48 + C.$$

Therefore, $C = 40$ and the position of the particle can be written

$$s(t) = \frac{5}{3}t^3 - 2t^2 + 7t - 40$$

9. C

Total distance traveled is given by $\int_0^2 |v(t)| dt$. On this interval, though, $v(t)$ is positive, so this is just

$$s(2) - s(0) = \left(\frac{5}{3}2^3 - 2 \cdot 2^2 + 7 \cdot 2 - 40\right) - \left(\frac{5}{3}0^3 - 2 \cdot 0^2 + 7 \cdot 0 - 40\right)$$
$$= \frac{40}{3} - 8 + 14$$
$$= \frac{58}{3}$$

10. C

The acceleration function is the second derivative of the position function, so

$$s(t) = \frac{1}{2}t^3 - 5t^2 + 3t + 6$$
$$s'(t) = \frac{3}{2}t^2 - 10t + 3$$
$$s''(t) = 3t - 10$$
$$a(t) = 3t - 10$$

To find the time when the acceleration is equal to -1 ft/sec^2, we solve:

$$a(t) = -1$$
$$3t - 10 = -1$$
$$3t = 9$$
$$t = 3 \text{ seconds}$$

Real-World Applications

WHAT ARE WORD PROBLEMS?

Calculus has many applications in real-life situations, including fields such as physics, chemistry, biology, business, economics, and the social sciences. These subject areas often appear in calculus as word problems. The key to successfully solving calculus word problems is being able to extract a function from the situation described in the problem and recognizing when to differentiate or integrate.

CONCEPTS TO HELP YOU

1. Rates of change: The derivative of a function is its rate of change, and lots of real-life applications of the derivative involve modeling functions' rates of change. These types of problems ask for the rate at which something increases or decreases.

2. Optimization: Optimization refers to finding the maximum or minimum value of a quantity, usually subject to specific conditions. If the quantity we are trying to optimize is described by a continuous function, we can use derivatives to find the maximum or minimum value of the function.

3. Related rates: We are often given information about the rate at which one quantity changes and then asked to use that information to provide answers about changes in another, related quantity. Because these types of problems involve two quantities, we use implicit differentiation to solve them.

4. Position, velocity, and acceleration: Velocity is the rate of change of position and acceleration is the rate of change of velocity. Inversely, velocity is the integral of acceleration and position is the integral of velocity.

5. Accumulated change: We can express the total change in a quantity over a time period as the definite integral of its rate of change over that time period. We can also use the definite integral to think about change thus far or to evaluate how much of a total quantity has accumulated at a certain point in time.

STEPS YOU NEED TO REMEMBER

1. *Identify what is given and what you need to find.*

Read the problem carefully and identify what answer you need as well as its units of measure. Single out the facts you need to answer the question. If possible, draw a picture of the problem.

2. *Extract a function from the description.*

Ask yourself "What computations must I do to solve this problem?" Depending on the information provided in the problem, you will either be differentiating or integrating.

3. *Write an equation to relate what is given to what you need to find.*

Set up an equation with the derivative or the integral and any known values.

4. *Solve the equation.*

Solve for the variable.

5. *Check the solution.*

Make sure the solution answers the question and makes sense in light of the facts presented in the problem.

COMMON WORD PROBLEMS

Rate of Change:

An 1800-gallon tank of water drains from the bottom in 30 minutes. The volume of water remaining in the tank after t minutes is

$$V = 1800\left(1 - \frac{t}{30}\right)^2, \text{ where } 0 \leq t \leq 30$$

How rapidly is the water draining from the tank after 15 minutes?

Step 1: Identify what you are given and what you need to solve.

Step 2: Extract the derivative that you need to solve.

Step 3: Substitute given values into the derivative to find the rate of change.

> **Solution and Explanation:** We know this is a rate of change problem because we are asked to find how fast (the rate) at which water is draining (volume decreasing). We are given information defining the volume of water in the tank at any particular time as $V = 1800\left(1 - \dfrac{t}{30}\right)^2$.
>
> The equation that we need to solve is $\dfrac{d}{dt}V(t)$ because the rate of change of the volume of water is given by the derivative.
>
> $$V'(t) = 3600\left(1 - \frac{t}{30}\right)\left(-\frac{1}{30}\right)$$
> $$= -120\left(1 - \frac{t}{30}\right)$$
>
> When $t = 15$
>
> $$V'(15) = -120\left(1 - \frac{15}{30}\right)$$
> $$= -60 \text{ gallons per minute}$$
>
> After 15 minutes, water is draining at the rate of 60 gallons per minute. The negative sign indicates that the volume of water is getting smaller.
>
> The next problem involves finding the maximum or minimim value of a function.

Optimization: Max or Min Value of a Function

(a) Find the dimensions of the pigpen that will minimize Farmer Fred's fencing cost.

(b) Find the cost of the fencing materials for that pigpen, to the nearest dollar.

(c) To the nearest decimeter, how much fencing material will Farmer Fred need to buy to construct the pigpen?

Step 1: Draw a picture.

Step 2: Write down the quantity to be maximized or minimized.

Step 3: Define variables; one variable should be the quantity you are trying to optimize.

Step 4: Write a function for the quantity to be optimized in terms of the other variables in the problem.

Step 5: Write the constraint equation(s) (if there are any) in terms of your variables.

Step 6: If the function you are trying to optimize is a function of more than one variable, use the constraint equation to eliminate all but one variable.

Step 7: Optimize the function by finding the critical points, determining the nature of each critical point, and evaluating the function at the critical points and at the endpoints of the domain (if applicable).

Step 8: Make sure to answer the question that is asked, paying attention to units and rounding.

Solution and Explanation: We are given that the fencing costs $20/meter and that the total area of the pigpen will be 150 square meters. We are trying to minimize the cost of the fencing materials for the pigpen. Let's start by drawing a picture.

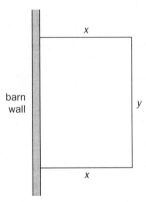

Looking at our figure, the total amount of fencing is the sum of the length of the three free-standing sides of the fence. Therefore, the cost C is $20 times the total length of fencing that needs to be purchased. We express this algebraically as $C = 20(2x + y)$.

We call the equation $C = 20(2x + y)$ the *equation to be optimized*. The minimum value of this equation can only occur at a critical point, which we can find by setting the derivative equal to zero. However, there are two variables in the equation. Before we find the critical points, we need to eliminate one of them. We eliminate either x or y by writing a *constraint equation* that allows us to express one variable in terms of the other.

We know that Farmer Fred needs the pigpen to have an area of 150 square meters. We can express this constraint as $xy = 150$. This is our constraint equation. When we solve the constraint equation for y, we get $y = \dfrac{150}{x}$. We substitute this into the equation to be optimized and get $C = 20\left(2x - \dfrac{150}{x}\right)$, where $x > 0$.

To find the critical points of C, we find its first derivative and set it equal to zero:

$$C' = 20\left(2 - \frac{150}{x^2}\right)$$

$$20\left(2 - \frac{150}{x^2}\right) = 0$$

$$2 - \frac{150}{x^2} = 0$$

$$2x^2 - 150 = 0$$

$$x^2 - 75 = 0$$

$$x^2 = 75$$

$$x = \pm 5\sqrt{3}$$

Since $x > 0$, the only critical point of C is $x = 5\sqrt{3}$. (Also, the length of x cannot possibly be $-5\sqrt{3}$.)

We must now verify that this critical point is in fact a minimum by using the Second Derivative Test. Recall that f has a relative minimum at $x = c$ if $f''(c) > 0$ and f has a relative maximum at $x = c$ if $f''(c) < 0$.

$$C'' = 20\left(\frac{300}{x^3}\right)$$

$$C''\left(5\sqrt{3}\right) = \frac{6000}{375\sqrt{3}} > 0$$

Because $C''\left(5\sqrt{3}\right) > 0$, it follows that C has a relative minimum at $x = 5\sqrt{3}$.

(a) We found that $x = 5\sqrt{3}$. Therefore, $y = \dfrac{150}{5\sqrt{3}} = 10\sqrt{3}$. The dimensions of the pigpen that minimize the cost of the fencing materials are $5\sqrt{3}$ m $\times$ $10\sqrt{3}$ m.

(b) We find the cost of the pigpen by plugging the dimensions of the pigpen into the cost equation:

$$C = 20\left(2x + y\right)$$
$$= 20\left(2 \cdot 5\sqrt{3} + 10\sqrt{3}\right)$$
$$= \$693$$

(c) To build the pigpen, Farmer Fred needs two sides of length x and one side of length y. That is, he needs $2 \cdot 5\sqrt{3} + 10\sqrt{3} = 34.6$ m of fencing material.

The following is an example of a related-rates problem.

Related Rates:

Claire is rollerblading due east toward the mall at a speed of 10 km/hr. John is biking due south away from the mall at a speed of 15 km/hr. Let x be the distance between Claire and the mall at time t and let y be the distance between John and the mall at time t.

(a) Find the distance, in kilometers, between Claire and John when $x = 5$ km and $y = 12$ km.

(b) Find the rate of change, in km/hr, of the distance between Claire and John when $x = 5$ km and $y = 12$ km.

(c) Let θ be the angle formed by the mall, Claire, and John (Claire is the vertex of the angle). Find the rate of change of θ, in radians per hour, when $x = 5$ km and $y = 12$ km.

Step 1: Draw a diagram.

Step 2: Identify the quantities that are changing and assign variables to represent them. Add labels to the diagram.

Step 3: Write down an equation that relates the quantities.

Step 4: Considering each variable as a function of t, use implicit differentiation to differentiate both sides of the equation with respect to t to find the rate of change (derivative) of each quantity.

Step 5: Restate the given problem(s) in terms of your variables and their rates of change (derivatives).

Step 6: Use Step 4 to solve the problem(s) in Step 5.

Solution and Explanation:

(a) Start by making a diagram to clarify the information presented in the problem:

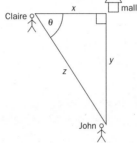

Let Claire's position be x, because Claire is traveling east. John's position can be y, because John is traveling south. We call the distance between Claire and John z. The diagram is a right triangle, so we can relate x, y, and z with the Pythagorean Theorem: $x^2 + y^2 = z^2$.

The problem tells us that $x = 5$ and $y = 12$. Using the Pythagorean Theorem, we find that

$$5^2 + 12^2 = z^2$$
$$169 = z^2$$
$$13 = z$$

The distance between John and Claire is 13 km.

(b) In part (a), we found an equation relating x, y, and z. To find an equation relating the velocities, we use implicit differentiation to differentiate the equation from part (a): $2x\dfrac{dx}{dt} + 2y\dfrac{dy}{dt} = 2z\dfrac{dz}{dt}$.

We are also given the information that Claire is traveling toward the mall at 10 km/hr. Her distance from the mall is decreasing, so $\dfrac{dx}{dt} = -10$.

We are told that John is traveling away from the mall at 15 km/hr. His distance from the mall is increasing, so $\frac{dy}{dt} = 15$. We are asked to find the rate of change for the distance between John and Claire when $x = 5$ and $y = 12$. We can cancel the twos in the equation above and fill in all of the required information.

$$x\frac{dx}{dt} + y\frac{dy}{dt} = z\frac{dz}{dt}$$

$$(5)(-10) + (12)(15) = 13\frac{dz}{dt}$$

$$130 = 13\frac{dz}{dt}$$

$$10 = \frac{dz}{dt}$$

The rate of change for the distance between Claire and John is 10 km/hr.

(c) Let θ be the angle formed by Claire's path and the hypotenuse of the right triangle. The rate of change of θ with respect to time is $\frac{d\theta}{dt}$. Using right triangle trigonometry, we know $\sin\theta = \frac{y}{z}$.

We differentiate this equation with respect to time. Notice that both y and z are variables with respect to time:

$$\cos\theta \cdot \frac{d\theta}{dt} = \frac{z \cdot \dfrac{dy}{dt} - y \cdot \dfrac{dz}{dt}}{z^2}$$

We can now substitute in the values used in part (b). Via right triangle trigonometry, we have $\cos\theta = \frac{x}{z} = \frac{5}{13}$. We find

$$\frac{5}{13} \cdot \frac{d\theta}{dt} = \frac{13(15) - 12 \cdot \dfrac{130}{13}}{13^2}$$

$$\frac{d\theta}{dt} = \frac{13}{5} \cdot \frac{13(15) - 12 \cdot \dfrac{130}{3}}{13^2}$$

$$= \frac{15}{13}$$

$$\approx 1.154 \text{ radians/hr}$$

The following is an example of a real-life situation involving position, velocity, and acceleration.

Postion, Velocity, and Acceleration:

An airplane's acceleration from the moment of lift-off $(t = 0)$ until 20 minutes into the flight is shown below.

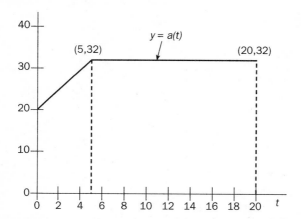

If the velocity at lift-off is 900 ft/min, what is the airplane's velocity after 20 minutes?

Step 1: Identify what is given and what you need to find.

Step 2: Decide whether to differentiate or integrate.

Step 3: Solve the derivative or integral.

Solution and Explanation: We are given the airplane's acceleration in the form of a curve from $t = 0$ to $t = 20$. We are also given the velocity at $t = 0$. The problem asks us to find the velocity at $t = 20$.

The airplane's velocity 20 minutes after lift-off is the initial velocity plus the change in velocity on $0 \le t \le 20$. The change in velocity is given by the definite integral $\int_{0}^{20} a(t)dt$. Therefore:

$$v(20) = v(0) + \int_{0}^{20} a(t)dt = 900 + \int_{0}^{20} a(t)dt$$

The definite integral $\int_{0}^{20} a(t)dt$ is the area under the acceleration curve on $0 \le t \le 20$.

We break this area into two regions: trapezoid A and trapezoid B:

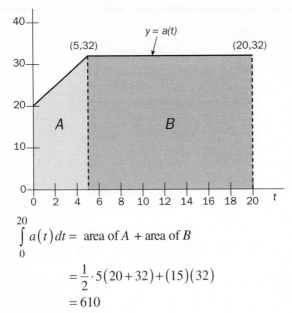

$$\int_0^{20} a(t)\,dt = \text{area of } A + \text{area of } B$$

$$= \frac{1}{2}\cdot 5\left(20+32\right)+\left(15\right)\left(32\right)$$

$$= 610$$

Therefore, the airplane's velocity at $t = 20$ is $v(20) = 900 + 610 = 1510$ ft/min.

The following is an example of an accumulated change problem.

Accumulated Change:

The volume of water in a tank changes at a rate given by the equation

$V'(t) = \dfrac{4t}{\sqrt[3]{t^2+3}}$, for $t \geq 0$. $V(t)$ has units in gallons and t has units in hours.

At $t = \sqrt{5}$ the tank holds 9 gallons.

What is the volume of water in the tank after 4 hours?

Step 1: Identify what you are given and what you need to find.

Step 2: Extract the definite integral that you need to solve.

Step 3: Substitute known values into the definite integral and solve.

Solution and Explanation: We are given the rate of change

of volume as $V'(t) = \dfrac{4t}{\sqrt[3]{t^2+3}}$ and asked to find how much has

accumulated in the tank $V(t)$ after a certain period of time.

The volume of water in the tank at time t is given by the expression $V(t) = \int V'(t)\,dt$.

Use u-substitution and manipulate constants to find an antiderivative for $V'(t)$:

$$u = t^2 + 3$$
$$du = 2t\ dt$$
$$2\,du = 4t\ dt$$

$$\int V'(t)\,dt = \int \frac{\overbrace{4\,dt}^{2\,du}}{\underbrace{\sqrt[3]{t^2+3}}_{u}} = \int \frac{2}{\sqrt[3]{u}}\,du = 2\int u^{-\frac{1}{3}} = 2\left(\frac{3}{2}u^{\frac{2}{3}}\right) = 3u^{\frac{2}{3}}$$

Substitute back into the antiderivative to get:

$$3u^{\frac{2}{3}} = 3\left(t^2+3\right)^{\frac{2}{3}} + C$$

To determine the constant C, plug in the initial condition: at $t = \sqrt{5}$, $V(t) = 9$.

$$3\left(\left(\sqrt{5}\right)^2 + 3\right)^{\frac{2}{3}} + C = 9$$

$$3(5+3)^{\frac{2}{3}} + C = 9$$

$$3(4) + C = 9$$

$$C = -3$$

So our antiderivative is $V(t) = 3\left(t^2+3\right)^{\frac{2}{3}} - 3$.

At $t = 4$, the volume of water in the tank is

$$V(4) = 3\left(4^2+3\right)^{\frac{2}{3}} - 3$$

$$= 3(19)^{\frac{2}{3}} - 3$$

CHAPTER QUIZ

1. The number of bacteria in a Petri dish after t hours is $n(t) = 3t^3 - 5t^2 + 2t - 2$. The population of baceria increases by a certain number of individuals per hour. How fast is the population growing after 4 hours?

   (A) 106

   (B) 85

   (C) 200

   (D) 70

   (E) 317

2. The value of a car after t years is $V(t) = 100t^2 - 300t + 150$ dollars; it depreciates at a rate of a certain number of dollars per year. At what rate does the car depreciate after 7 years?

   (A) $2000

   (B) $1000

   (C) $1100

   (D) $550

   (E) $725

3. An apple farmer can expect 600 apples from each of his apple trees if no more than 20 trees are planted. If he plants more than 20 trees, the yield per tree will decrease. For each extra tree he plants, his yield per tree will decrease by 15 apples. How many trees should he plant to obtain the maximum number of apples?

   (A) 450

   (B) 40

   (C) 200

   (D) 10

   (E) 30

4. The wholesale price of a designer shirt is $25. If a retail store sells the shirt for $40, they will sell 55 shirts per month. For every dollar that the store lowers the shirt's price, 5 more shirts will be sold each month. What selling price will yield the greatest monthly profit for the store?

   (A) $39

   (B) $38

   (C) $37

   (D) $36

   (E) $35

5. The lengths of the sides of a square are decreasing at a constant rate of 4 ft/min. In terms of the perimeter P, what is the rate of change for the area of the square in square feet per minute?

   (A) $-2P$

   (B) $2P$

   (C) $-4P$

   (D) $4P$

   (E) $8P$

6. The base of a triangle is decreasing at a constant rate of 0.2 cm/sec and its height is increasing at a rate of 0.1 cm/sec. If the area is increasing, which answer best describes the constraints on the height h at the instant when the base is 3 cm?

   (A) $h > 3$

   (B) $h < 1$

   (C) $h > 1.5$

   (D) $h < 1.5$

   (E) $h > 2$

7. You are flying from Chicago to Atlanta. Shortly after takeoff, the captain turns off the "fasten seatbelts" sign and announces that the plane has now reached a cruising speed of 635 miles per hour. Given that velocity is the derivative of position and that acceleration is the derivative of velocity, what is the acceleration of the plane at that moment?

   (A) 200 mi/hr^2

   (B) 100 mi/hr^2

   (C) 3300 mi/hr^2

   (D) -635 mi/hr^2

   (E) None of the above

8. The graph below shows the velocity of two bicyclists, *A* and *B*, in a race from time $t = 0$ to time $t = 16$. They start side by side and travel along the same road. The race begins at time $t = 0$, and at time $t = 16$, one of the racers completes the course.

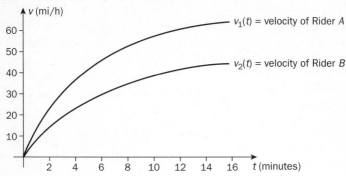

What is the total length of the course?

(A) $\int_0^{16} v_1(t)\,dt$

(B) $\int_0^{16} v_2(t)\,dt$

(C) $\int v_1(t)\,dt$

(D) $\int v_2(t)\,dt$

(E) Cannot be determined

9. A particle is moving along the x-axis with velocity
$v(t) = \sin t + \cos t$, for $t \geq 0$.
What is its maximum acceleration over the interval $[0, 2\pi]$?

(A) $\cos \dfrac{3\pi}{4} + \sin \dfrac{3\pi}{4}$

(B) $\cos 2\pi + \sin 2\pi$

(C) $\cos \dfrac{7\pi}{4} + \sin \dfrac{7\pi}{4}$

(D) $\cos \dfrac{7\pi}{4} - \sin \dfrac{7\pi}{4}$

(E) $\cos 0 + \sin 0$

10. Water flows into a pond at a rate of $300\sqrt{t}$ gallons/hour and flows out at a rate of 400 gallons/hour. After 1 hour there are 10,000 gallons of water in the pond. How much water is in the pond after 9 hours?

(A) 10,000

(B) 11,000

(C) 12,000

(D) 14,000

(E) 16,000

Answers and Explanations

1.   A

We are given the equation for the number of bacteria in the Petri dish and asked to determine how fast (the rate) the population is growing after 4 hours.

Because the number of bacteria is given by $n(t) = 3t^3 - 5t^2 + 2t - 2$, the rate at which the number changes is given by:

$$n'(t) = 9t^2 - 10t + 2$$

At $t = 4$ hours

$$n'(4) = 9(4)^2 - 10(4) + 2$$
$$= 106$$

After 4 hours, the bacteria are growing at the rate of 106 per hour.

2.   C

We are given the value of a car as $V(t) = 100t^2 - 300t + 150$ dollars and asked to find the rate at which the car's value depreciates.

The rate at which the car depreciates is given by

$$V'(t) = 200t - 300$$

After 7 years

$$V'(7) = 200(7) - 300$$
$$= 1100$$

After 7 years, the car depreciates at the rate of $1100 per year.

3.   E

Let x = number of additional trees that the farmer plants (beyond 20), so that the total number of trees planted is $x + 20$. Then, the yield of apples per tree is $600 - 15x$. The total number of apples that can be picked, T, is the product of these two values:

$$T = (x + 20)(600 - 15x) = 600x - 15x^2 + 1200 - 300x = 300x - 15x^2 + 1200.$$

We want to find the maximum value of T, which requires that we compute T' and set it equal to zero: $T' = 300 - 30x = 0 \Rightarrow x = 10$ additional trees planted.

Thus the number of trees the farmer should plant is $x + 20 = 30$ trees. Note that the second derivative is -30, which is negative. Therefore this amount is a maximum.

4.   B

Let x = the *decrease* in price per shirt.

The selling price is therefore $40 - x$ dollars.

The total profit is $P(x)$ = (profit per shirt) $\times$ (number of shirts sold).

The store's profit per shirt is the wholesale price subtracted from the retail price:

$$(40 - x) - 25 = 15 - x$$

Because 5 more shirts are sold for each dollar decrease in price, the number of shirts sold per month is $55 + 5x$.

Therefore, the profit is

$$P(x) = (15 - x)(55 + 5x)$$
$$= 825 + 20x - 5x^2$$

To find the maximum profit, we differentiate $P(x)$ and set it equal to zero:

$$P'(x) = 20 - 10x$$
$$20 - 10x = 0$$
$$-10x = -20$$
$$x = 2$$

The second derivative is -10, so $x = 2$ represents a relative maximum. Since there is only one relative extremum, the absolute maximum profit occurs when $x = 2$.

The optimal selling price is $\$40 - \$2 = \$38$.

5.　A

This is a related-rates question, with a twist. The problem gives us information about the rate of change of the side of the square; we then need to answer a question about the rate of change for the area. After we do this, we are done with the related rates part of the problem and must express our answer in terms of the perimeter.

Following the procedure for related-rates problems, begin by drawing a diagram and identifying the quantities in the problem.

We find the area of the square in terms of our variables $A = s^2$. We differentiate both sides of this equation with respect to t to find an equation relating $\frac{dA}{dt}$ and $\frac{ds}{dt}$: $\underbrace{\frac{dA}{dt} = 2s\frac{ds}{dt}}_{\text{chain rule}}$. The problem tells us that the length of the square's sides is decreasing at a rate of 4 ft/min, i.e., $\frac{ds}{dt} = -4$. We substitute this information into the equation for $\frac{dA}{dt}$ above and find $\frac{dA}{dt} = 2s \cdot -4 = -8s$.

To complete the problem, we need to express $\frac{dA}{dt}$ in terms of the perimeter. The perimeter of a square is $P = 4s$. Therefore $s = \frac{P}{4}$. Substituting this expression for s into the expression above, we find

$$\frac{dA}{dt} = -8 \cdot \frac{P}{4} = -2P. \text{ Therefore } \frac{dA}{dt} = -8s = -8\left(\frac{P}{4}\right) = -2P.$$

6.　D

Following the procedure for related-rates problems, first draw a diagram and label the quantities in the problem.

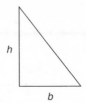

Next, find an equation relating the quantities in the problem—in this case, the formula for the area of a triangle: $A = \frac{1}{2}bh$. We differentiate both sides of the equation with respect to t using the product rule: $A'(t) = \frac{dA}{dt} = \frac{1}{2}\left(b\frac{dh}{dt} + h\frac{db}{dt}\right)$. We express the information provided in the problem in terms of our variables: $\frac{db}{dt} = -0.2$ cm/sec, $\frac{dh}{dt} = 0.1$ cm/sec, $b = 3$cm and substitute the information into the expression above, giving $A = \frac{1}{2}(3(+0.1) + h(-0.2))$.

The problem tells us that the area is increasing, i.e., $A'(t) > 0$. Therefore we should set the expression for A' above greater than zero and solve for h:

$$\frac{1}{2}(3\cdot(+0.1) + h\cdot(-0.2)) > 0 \quad \Rightarrow \quad h\cdot(-0.2) > -0.3$$

$$h\cdot(-0.2) > -0.3 \quad \Rightarrow h < 1.5$$

7. E

When the velocity is constant, the rate of change of the velocity—which is the acceleration—is zero.

8. A

The area under each curve represents the total distance traveled by the respective racers from $t = 0$ to $t = 16$; the area between the curves represents the distance between the riders at time $t = 16$.

Because the area under $v_1(t)$ is greater than the area under $v_2(t)$, A has traveled farther at $t = 16$. Therefore, this rider has completed the course and the total length of the course is given by the definite integral $\int_0^{16} v_1(t)\,dt$.

9.   D

Acceleration is the derivative of velocity, so to find an equation for the acceleration, we differentiate the velocity function

$$v(t) = \sin t + \cos t \quad \Rightarrow \quad v'(t) = a(t) = \cos t - \sin t$$

Because the acceleration function is continuous, the maximum acceleration can only occur at the critical points (where $a'(t)$ is zero or undefined) or at one of the endpoints of the interval $[0, 2\pi]$.

$$a(t) = \cos t - \sin t \quad \Rightarrow \quad a'(t) = -\sin t - \cos t$$

$$a'(t) = -\sin t - \cos t = 0 \quad \Rightarrow \quad \sin t = -\cos t \quad \Rightarrow \quad \tan t = -1$$

$\tan t = -1$ in the second and fourth quadrants at $t = \dfrac{3\pi}{4}, \dfrac{7\pi}{4}$

Notice that the derivative $a'(t)$ is defined everywhere, so the only critical points are the points at which $a'(t) = 0$. We can make a table of the critical points and endpoints and the value of the acceleration $a(t)$ at each of these points.

t	$\dfrac{3\pi}{4}$	$\dfrac{7\pi}{4}$	0	2π
$a(t)$	$\cos\dfrac{3\pi}{4} - \sin\dfrac{3\pi}{4} =$ $-\dfrac{\sqrt{2}}{2} - \dfrac{\sqrt{2}}{2} = -\sqrt{2} \approx -1.414$	$\cos\dfrac{7\pi}{4} - \sin\dfrac{7\pi}{4} =$ $\dfrac{\sqrt{2}}{2} - \left(-\dfrac{\sqrt{2}}{2}\right) \approx 1.414$	$\cos 0 + \sin 0$ $= 1 + 0$ $= 1$	$\cos 2\pi + \sin 2\pi$ $= 1 + 0$ $= 1$

Of these four choices, the maximum acceleration is $\sqrt{2}$; we are guaranteed that the maximum acceleration is one of these four values, so the correct answer is (D).

10. C

The total rate of change of the volume of water in the pond is
$r(t) = $ *rate in − rate out.*

We are given that the rate in is $300\sqrt{t}$ gallons/hr and the rate out is 400 gallons per hour. Therefore $r(t) = 300\sqrt{t} - 400$.

If we let $A(t)$ be the amount of water in the pond at time t, then A is an antiderivative of r. Let's approach this problem as an accumulation function. Because we are given the volume of water at $t = 1$ and we are asked to find the volume of water at $t = 9$, we can set the problem up as $A(9) = A(1) + $ *change in volume on* $1 \le t \le 9$. The change in the volume of water is given by the definite integral $\int_1^9 r(t)\, dt = \int_1^9 300\sqrt{t} - 400\, dt$, so

$$A(9) = 10,000 + \int_1^9 300\sqrt{t} - 400\, dt$$

$$= 10,000 + \int_1^9 300t^{\frac{1}{2}} - 400\, dt$$

$$= 10,000 + \left(\frac{2}{3} \cdot 300 t^{\frac{3}{2}} - 400t \right)\Bigg]_1^9$$

$$= 10,000 + \left(200 t^{\frac{3}{2}} - 400t \right)\Bigg]_1^9$$

$$= 10,000 + \left(200 \cdot 9^{\frac{3}{2}} - 400 \cdot 9 \right) - \left(200 \cdot 1^{\frac{3}{2}} - 400 \cdot 1 \right) = 12,000$$